Building Materials and Technology in Hong Kong
香港建築技術及應用

Building Materials and Technology in Hong Kong

香港建築技術及應用

Wong Wah Sang 黃華生
Chan Wing Yan, Alice 陳詠欣
Wai Chui Chi, Rosman 衞翠芷
Kee Yee Chun, Tris 祁宜臻

Hong Kong University Press
The University of Hong Kong
Pokfulam Road
Hong Kong
www.hkupress.hku.hk

© 2018 Hong Kong University Press

ISBN 978-988-8390-98-4 (*Hardback*)

All rights reserved. No portion of this publication may be reproduced or transmitted in any form or by any means, electronic or mechanical, including photocopy, recording, or any information storage or retrieval system, without prior permission in writing from the publisher.

British Library Cataloguing-in-Publication Data
A catalogue record for this book is available from the British Library.

10 9 8 7 6 5 4 3 2 1

Printed and bound by Hang Tai Printing Co. Ltd., Hong Kong, China

Contents
目 錄

Foreword by Vincent Ng 序　吳永順	viii
Preface by Wong Wah Sang 前言　黃華生	ix
Preface by Chan Wing Yan, Alice 前言　陳詠欣	xi
Acknowledgements by Wong Wah Sang 鳴謝　黃華生	xii

1. **General Introduction to Building Construction in Hong Kong** — 1
   Wong Wah Sang
   香港建築簡介　黃華生

2. **Trades of Materials and Technology** — 11
   Wong Wah Sang
   工科的類別及技術　黃華生

   2.1　Preliminaries — 12
   　　　整體施工預備工作

   2.2　Demolition and Excavation Works — 22
   　　　拆樓及挖掘工程

   2.3　Concrete Work — 27
   　　　混凝土工作

   2.4　An Outline of Foundation Systems in Hong Kong — 44
   　　　香港常見的地基種類

   2.5　Brickwork and Blockwork — 49
   　　　磚塊

   2.6　Masonry and Granite/Marble Works — 56
   　　　石工、雲石和麻石

   2.7　Roofing, Waterproofing, and Expansion Joints — 64
   　　　屋頂的鋪工、防水層、伸縮縫

   2.8　Carpentry, Joinery, and Ironmongery — 77
   　　　木工、細木工、五金

   2.9　Staircases, Steps, and Handrails — 87
   　　　梯級及欄杆扶手

2.10 Metal Windows and Doors 92
金屬門窗

2.11 Glazing, Curtain Wall, and Cladding 97
裝配玻璃、玻璃幕牆與外牆板模

2.12 Floors, Walls, and Ceiling Finishes 112
地台、牆身、天花物料

2.13 Plasterwork 122
抹灰工程

2.14 Painting 130
油漆工程

2.15 Builders' Work in Relation to Plumbing, Drainage, and Mechanical and 134
Electrical Services
有關渠道、水喉、機電工程的建築項目

2.16 External Work and Landscape Work 148
外圍工作與園境計劃

3. **Case Studies** 155
Wong Wah Sang and Chan Wing Yan, Alice
實例研究　黃華生、陳詠欣

3.1 The Forum, Hong Kong, China: Curtain Wall Design 156
中國香港富臨閣：玻璃幕牆設計

3.2 CIC Zero Carbon Building: Eco-Building Design and Technologies 163
零碳天地：環保建築及技術

3.3 Domain: Redevelopment of Yau Tong Estate Phase 4, a Sustainable 169
Commercial Building
「大本型」：油塘邨第四期重建發展——推動可持續發展的商業建築

3.4 Harmony 2, Tin Yiu Estate, Phase 3, Tin Shui Wai: Prefabrication Works 174
天水圍天耀邨和諧式的設計組合及預製模件

3.5 Residences at Nos. 96, 98, and 100, Ma Ling Path, Sha Tin: 186
Single-Storey House Construction
沙田馬鈴徑洋房：小型獨立式住宅建築

3.6 Mong Tung Wan Youth Hostel: Small Building Construction 195
望東灣青年旅舍：小型建築建設技術

3.7 Birchwood Place, 96 MacDonnell Road: High-Rise Residential Building 202
on Slope
麥當奴道96號寶樺臺：斜坡上的高廈建設

3.8 The French International School: External Wall and Auditorium 214
法國國際學校：外牆與禮堂建設

3.9 The Heungs' Residence at the Peak: Granite and Glass Technology 221
山頂香氏大宅：麻石與玻璃裝置技術

3.10 The St. John's Building, Garden Road: Curtain Wall and Cladding 234
花園道聖約翰大廈：玻璃幕牆及覆蓋板

3.11	Wanchai Indoor Games Hall: Roof Truss Construction 灣仔室內運動場：樑架建築樣式	240
3.12	Printing House Vertical Extension: Building on Top of an Existing Building 印刷行大廈：樓層之上的擴建工程	246
3.13	Hong Kong Science Museum: Cavity Wall and External Works 香港科學館：空心牆與外圍工程	258
3.14	Citibank Plaza, Garden Road: An Intelligent Building 萬國寶通廣場：智慧型大廈	270
3.15	Central Plaza: Super High-Rise Concreting Technology 中環廣場：超級高廈的混凝土建築技術	282
3.16	Sam Tung Uk Museum: Landscaping and External Works 三棟屋博物館：園林設計	295

4. Drawing Practices: From Design Sketches to Tender Drawings 303
Kee Yee Chun, Tris
從草圖到招標圖　祁宜臻

5. The Importance of Construction Specifications 333
Wai Chui Chi, Rosman
施工規格的重要性　衛翠芷

About the Authors 343

Foreword

The mission of the Hong Kong Institute of Architects is to promote the general advance of architecture and to promote and facilitate the acquisition of the knowledge of the various arts and science connected therewith. The Institute also strives to raise the standard of architecture in Hong Kong and of professional architectural services offered by its members. I have found the purpose of this book aligned with the mission of the Institute.

As architects, we design building forms with respect to function, to fulfil the need of users, and to comply with the current building laws and codes. Further, it is essential for architects to be equipped with an understanding of building components and the knowledge of building materials and technology. It is often with this knowledge of how buildings are assembled that innovative ideas are materialized.

Construction technology is a subject undergoing continuous development, directly or indirectly influenced by technological advancement, people's aspiration for sustainability, changes in building laws and codes, and local market conditions. I am therefore pleased to note that this edition has been updated with case studies of recent local projects, such as CIC Zero Carbon Building and domain-redevelopment of Yau Tong Estate Phase 4.

I commend the lead author, Dr Wong Wah Sang, a renowned architect and educator, for his commitment, persistence, and unfailing efforts in sharing his valuable research with his co-authors in the profession. I am pleased to recommend this book as a truly indispensable resource to all architectural students and practitioners.

Vincent NG, JP
President
The Hong Kong Institute of Architects
November 2015

Preface

First published in 1991 and updated in 2017, this book is a record of three decades of building materials and technology in Hong Kong. It is still a valuable reference for looking at how buildings have been designed and built in a high-density city in a subtropical climate. Advances in technology and more concern about environmental awareness have contributed to variation in design and construction. Increased density and complexity in statutory constraints have not stopped innovation in architecture, and more buildings with a variety of interests and focus have appeared recent years. When we examine the details and the buildings, we see that fine detailing, good work, and efficient management are of no less importance than is the philosophy of architectural design in quality architecture.

This book is a thorough documentation of tectonics in the Hong Kong construction industry. After the introduction, Chapter 2, 'Trades of Materials and Technology' describes the trades of the building industry in various aspects of interest to an account of current construction standards in Hong Kong. In the last three decades there has been no new trade although materials have improved and have new attributes such as environmental responsiveness. However, the cost index has risen almost threefold in the past 25 years.

Reference is made to the relevant British standards and building regulations, quoting pertinent experience from construction sites. Chapter 2 is discussed mainly from the designer's point of view without highly technical details or experimental testing, which are left for the reader's own exploration. Nevertheless, construction trades can be better appreciated on site.

Chapter 3, 'Case Studies', includes a large variety of buildings for review. Particular emphasis on construction is discussed in individual cases—from single-storey buildings to high-rise towers, from private development to public institutional construction, from office to residence. Complications and specialty of construction are mentioned in these cases. Though these studies cannot claim to represent all building details in Hong Kong, the general construction experience over three decades in a high-density subtropical environment can be appreciated and shared. It is also noted that there has been continuous export of Hong Kong tectonics and practices into mainland China and Southeast Asia.

There are two additional chapters: Chapter 4, 'Drawing Practices: From Design Sketches to Tender Drawings' and Chapter 5, 'The Importance of Construction Specifications'. The former explains how the architect develops the sketches into a set of construction drawings that can be communicated to the contractor for constructing the building. The latter describes another important role of the architect as specification writer to establish the standard of construction for high-quality work. Both chapters are unique and among the first of their kind to appear in the literature of the Hong Kong building industry.

I would like to thank all the people whose various contributions have made this publication possible. Thanks to my co-authors, Alice Chan, Tris Kee, and Rosman Wai, who have contributed their time and shared their knowledge to enrich the contents of this book. Special thanks are due to Alice Chan, who initiated the updating and thus this new edition.

Finally, it is hoped that, besides serving as a reference for other professionals and students, this book will inspire designers and administrators in the building industry to create innovative quality buildings for the excellence of architecture.

Wong Wah Sang
2017

Preface

One of my life missions is to pass the architectural and building torch to the next generation. As a Hong Kong–based architect, from study to graduation, from practice to certification, I want to give back the things to society that I have taken up so far.

The universe of architecture and building technology is so broad that there is no all-round shortcut to learning comprehensive site matters; a consolidation of piece-by-piece on-the-job training and experience is required. I recall that when I first encountered a curtain-wall building design, I could scarcely find a reference book about building technology used in Hong Kong with real-life case studies. With reference to numerous engineering books as well as plenty of consultations with the façade consultants, I eventually mastered how the system works and how to design for the best. There is no doubt that a good reference book facilitates our learning and application. I would say that this updated edition is an excellent reference book for architects in Hong Kong. Therefore, I volunteered to be the editor for the new edition.

In this new edition, new inventions and technologies have been incorporated in addition to more practical knowledge and real-life experience. Five years of editing and preparing the case studies were hard work yet worthwhile. By connecting technical knowledge with real-life application, this book will continue its mission to inspire students of Hong Kong architectural history and technology, impart more architectural knowledge to a wider public, correct certain misconceptions, and raise public awareness of our environment. By passing the torch to the next generation, we look forward to a new era for architecture.

Chan Wing Yan, Alice
2017

Acknowledgements

I would like to acknowledge with great appreciation all the help I received from numerous developers, architects, engineers, and contractors who contributed to this book. The affiliated companies are listed at the time of the corresponding contribution). For their support of the case studies, many thanks are due to

> Mr Donald Choi of Sino Realty & Enterprises Limited,
> Mr Heung Chit Kau,
> Mr Barrie Ho of BARRIE HO Architecture Interiors Ltd.,
> Mr Dennis Lau Wing Kwong of Ng Chun Man & Associates Architects & Engineers (H.K.) Limited,
> Mr Patrick Lau of Design Consultants,
> Mr Stephen Lau of Design Design Architects,
> Mr Bernard Lim of P & T Architects,
> Mr Anthony Ng of Anthony Ng Architects,
> Mr S. H. Pau and Mr C. Y. Choi of the Architectural Services Department,
> Mr Philip So of Lee and So and Associates,
> Mr Tong Chun Wan of Great Eagle Project Management Limited,
> Mr Stephen K. K. Tong of the Housing Authority,
> Mr C. K. Wong, president of Hong Kong Institute of Landscape Architects,
> Mr Rocco Yim of Rocco Design Partners,
> Mr Jerome Wong, executive director of Aedas Limited,
> Mr M. K. Leung, director of Sustainable Design/associate director of Ronald Lu & Partners,
> Mr Tony Ip, associate/deputy director of Sustainable Design of Ronald Lu & Partners, and
> Mr Antony Chung of the Housing Authority.

For the arrangement of site visits, I would like to thank

> Mr David Chan of Hien Lee Engineering Company Limited,
> Mr C. Y. Choi of the Architectural Services Department,
> Mr Donald Choi of Sino Realty & Enterprises Limited,
> Mr Lee Yung Wong of Youth Hostel Association,
> Mr Bernard Lim of P & T Architects,
> Mr Tang Kar Hung of Tak Son Engineering Company Limited,
> Mr Tong Chun Wan of Great Eagle Project Management Limited,
> Mr Stephen K. K. Tong of the Housing Authority, and
> Mr Tso of the French International School.

For interviewing, writing, and gathering information for case studies,

> Miss Julie Cheng of the Architectural Services Department.

I would like to thank Miss Asuki Chan, Miss Ava Wang, Mr Samuel Cham, and Mr Leung Yat Hei of the Architectural Project Unit Limited for supplying new illustrations and editing the text.

I would also like to thank Mr Man Kwok Hung for his advice on the foundation systems in Hong Kong, Ms Vivian Ai for her copyediting, and Mr Yeung Tim Cheung for drafting the illustrations. Of course, special thanks go to my wife, Angela, for all her typing work.

There are also contributions directly or indirectly from a lot of people such as design and construction teams. Without their work and continuous development of new materials and technology in Hong Kong, we would not have so much information to publish in this book. I would like to express my gratefulness to them all.

Wong Wah Sang
2017

Chapter 1

General Introduction to Building Construction in Hong Kong

香港建築簡介

Wong Wah Sang 黃華生

The Bank of China standing amid shorter office buildings

Much as the trapezoidal shape of the East Building's site had generated its triangular design motif, the compelling diagonals of the Bank of China were intimately related to its aesthetics. Rather than leave the roofs of the ascending shafts flat, as might have been expected, Pei set them at an angle, adding to the sense of thrust as the building rose. The skin was to be reflective glass. In true Modernist tradition, Pei chose to express the structural members that met the skin by highlighting them with aluminium cladding, creating facades of boxed Xs. At seventy stories, the tower would be the tallest building in the world, outside the United States, and its strikingly abstract form—topped with a pair of broadcast masts—would make a dramatic vertical gesture in the heart of the city.

—Carter Wiseman, *I. M. Pei: A Profile in American Architecture*

In the late 1980s, people working in Central witnessed the construction of I. M. Pei's Bank of China Building in the non-Hong Kong-style construction method of using structural steel. This building with the 'trunk of a bamboo' metaphor is an example of continuous new technology being brought into the building industry, giving inspiration for Hong Kong architects. The 370 m height of the building is recorded as the highest in Hong Kong, offering a challenge to local architects.

Subsequently in 1992, the completion of Central Plaza, 18 Harbour Road, another corporate-looking office building became the tallest building in Hong Kong. And in 2010, another record was set by the 108-storey International Commerce Centre, 484 m high. Such are the continuous changes and impact that make Hong Kong construction so dynamic and interesting.

運用三角的意念於中國銀行大廈的設計，正好和斜形支架結構作了個和諧的組合。同時擺脫以往平平坦坦的頂部設計，貝聿銘的建築物造型配合了不同的取向和角度，並且在每個平面之外，披上一身反光的玻璃幕牆。在現代主義的傳統影響下，貝聿銘完全將建築物的結構顯露出來，利用鋁合金板和玻璃幕牆，成為整體的盒子形設計。整座建築物共七十層高，將會是美國之外最高的建築物了。超創意的靈感想像，還有，它在屋頂的一對巨型天線，成為市中心戲劇性的直立姿態。

—Carter Wiseman, *I. M. Pei: A Profile in American Architecture*

1980 年後期，我們在中環看到中國銀行大廈的建築過程——承重鋼鐵的結構建築法，並非一般香港式的釘板模、紮鐵、混凝土的方法。這比喻為「竹節」的建築，將新科技引進，也給予香港建築業一個嶄新的啟示和開始。全幢大廈高 370 米，是高廈設計上的一個新指標。其後 1992 年在灣仔海旁落成的中環廣場，更成為另一項高廈的紀錄。再後 2010 年，108 層的環球貿易廣場，高 484 米，創下另一香港高廈的紀錄。

Construction speed

Construction speed is of prime importance in the Hong Kong building process, even outweighing quality. 'Time is money' is the slogan for the rich developers who are continuously inserting buildings to reshape Hong Kong's skyline. The amazing speed of construction is as fast as three to four days per concreting cycle for a typical floor. Large housing estates in Laguna City of Cha Kwo Ling and the residential towers in Kingswood Estate, Tin Shui Wai, at a four-day

4 General Introduction to Building Construction in Hong Kong

Construction of the steel frame for the Bank of China

Office building at no. 9 Queen's Road

Citibank Plaza at a three-day cycle per floor

New Hang Seng Bank headquarters

Central Plaza, the tallest office building located at the Wanchai waterfront, under construction

Residential tower at Bonham Road

cycle are good examples of local construction speed. Citibank Plaza in Central went up at a three-day cycle.

A review of the high cost of land explains the emphasis on time. For example, the particular site in Wanchai for the tallest office building was acquired at a government auction on 25 January 1989, at HK$3,350,000,000 or HK$460,000 per m² site area. Interest on the cost of the land alone amounted to HK$918,000 per day at the prime rate of 10%. This would be enough pressure to produce building very fast (at a four-day cycle per floor). Subsequently in 2016, a business site, NKIL no. 6505, 7728 m², at Cheung Sha Wan, was sold at HK$7,794.38 million, about HK$1 million per m².

Small site at Queen's Road Central with narrow frontage, equivalent to the width of a van

建築工程速度

在香港，建築工程中最重要的考慮因素是時間，其次才談到質素與品質標準。俗語云：「時間即是金錢」，是許多發展商的金句，他們不斷在香港擠壓出高廈，重組香港的天際線。一般來説，普通的一層樓面建築，從釘板到澆灌混凝土，可以快至三至四天的時間，就以大型屋邨如茶果嶺的麗港城、天水圍的嘉湖山莊等，都是以四天的建築週期速度興建的。在中區的萬國寶通廣場甚至可達成三天的建築週期。高地價正好解釋時間的重要性。例如在灣仔1989年買入的一幅地的地價為三十三億五千萬元或每平方米四十六萬元。地價的息口計算已為每日九十一萬八千萬元。這因素已足夠壓力去推動一「速成」建築。及後2016年，一幅在長沙灣的商業用地，NKIL no. 6505, 7728平方米，以約七十八億元成交，每平方米地價為一百萬元。

Statutory restraints

Again due to the high land prices, developers and architects have constantly worked their way through legal restrictions (basically the Building Ordinance) to fish for possible developable areas. As example is the addition of top floors to the Sun Hung Kai Centre in Wanchai. This existing 42-storey building has been served by pedestrian walkways at the first-floor level. With precedent reference projects, application to the Building Ordinance Office was made for the dedication to the public of these first-floor pedestrian walkways and some ground floor public areas. The application was approved, and consequently a bonus floor area was added as the top seven floors to the existing building. A new category of design was opened up: the construction of a structure on top of an existing building when the building was still in operation. Another interesting phenomenon for this project is the conversion of existing lift shafts to accommodate double-decker lifts to meet increased passenger loading.

Just as difficult with large sites are small sites within the Central business district, say, the sort of 'one number' sites at Queen's Road Central.

Vertical extension at Sun Hung Kai Centre

法例規限

由於香港的土地價值昂貴，發展商都在法例規限的容許下，爭取更多的發展面積。這裏不妨以灣仔海旁的新鴻基中心作例子。發展商根據附近的同類發展，再向屋宇地政署申請，重新把一樓的行人天橋界定為公眾地方。而根據發展準則的案例，業主獲批准增加可發展的面積。這樣的

Kingswood Estate in Tin Shui Wai

發展或許可帶來新的設計方案。然而在四十二層樓高之上再加七層的面積，建築師需特別設計雙層電梯的方案來應付增加的運輸量。面積較小的發展地盤，比許多大規模的地盤施工更困難。沿著香港皇后大道中，便有著許多窄小而且只有一個地段編號的地盤。

Tradition

Though technology may be more advanced, and buildings are designed to be more 'intelligent', the average Hong Kong building still employs a lot of labour-intensive trades. Bamboo scaffolding, plywood formwork, spatterdash on concrete surfaces, cutting of reinforcement bars, sawing of timber planks, spray painting, hanging up mosaic tiles, etc. all are familiar scenes on a Hong Kong construction site. Similar techniques are observed in construction sites in South-East Asia, such as Singapore and Malaysia.

Not to be separated from the tradition is Lu Pan (魯班), the patron saint of Chinese builders and contractors. His festival, the 13th day of the 6th month in the Chinese calendar, is always marked with celebrations—dining, drinking, and gambling by the contractors. A tribute can be made to Lu Pan Temple, located in Ching Lin Terrace at Kennedy Town on Hong Kong Island.

傳統

雖然現代科技不斷改進和發展，東南亞區如星加坡、馬來西亞和香港的建築業，仍以勞動工人為主。故此，在一般香港建築地盤，仍然可以見到如蓋搭竹棚、釘木板、撒沙、拆鐵板、鋸木、噴漆、鋪砌紙皮石等工序。魯班被譽為中國工匠師祖。每年農曆六月十三日，所有建築從業員都會舉行特別的慶祝活動。而位於堅尼地城青蓮臺就建立了一座魯班先師廟。

Cost of skilled labour in March 2016 (information extracted from Wages and Labour Costs Statistics Section (1), Census and Statistics Department)
2016年3月的專業技術工資表 (資料摘取自政府統計處工資及勞工成本統計組 [一])

Trade (行業)	Cost per day (HK$) (每日工資港元$)
Setting out (開線、測量)	1399.1
Formwork (木板)	1949.6
Steelwork (紮鐵)	1996.5
Concreting (落石屎工)	1902.25
Plaster, tiling (泥水、磚瓦)	1221.4
Carpenter (木匠)	1192.4
Painting (油漆)	1141
Scaffolding worker (搭棚工人)	1748.8
Unskilled labour (非專業技工)	920.9

Celebrations at the Lu Pan Temple on 13th day of the 6th moon: the Lu Pan Festival

Lu Pan Temple

Labour for the construction industry

Traditionally, construction workers are mainly from China. Some work, such as reinforcement work, requires young and strong, while other work, such as joinery and painting, require older and more experienced workers.

The Construction Industry Training Authority (CITA), established in 1975, was amalgamated with the Construction Industry Council (CIC) in 2007. The functions of CITA in training local workers has been taken over by the Construction Industry Training Board (CITB) under CIC.

Basic craft courses in the CIC Training Academy include:

Two-year courses:
    1. Bricklaying, plastering, and tiling
    2. Carpentry and joinery
    3. Painting, decorating, and sign-writing
    4. Plumbing and pipe-fitting
    5. Marble-laying
    6 Metal work

One-year courses:
    1. Construction scaffolding work
    2. Construction plant maintenance and repairs
    3. Electrical installation

There is practical training in workshops and classroom lectures for trainees, without admission and tuition charges. During training, each trainee receives a basic monthly allowance of around HK$1,000. There is also a special allowance for site practice, in order to attract people to the construction industry.

Training Centre at Aberdeen with brickwork, plasterwork, and carpentry

建築工程資源

一般建築工程項目中需要勞動的工種如紮鐵等，都由年青和健壯的工人負責；而部份講求技術的工種（諸如木工、油漆等），則多半是以中年且經驗豐富的工人為主。除了由中國內地輸入的勞工之外，建造業訓練局於 1975 年成立，2008 年該局合併至建造業議會，其下設的建造業訓練委員會負責培訓本地建築工程的專門技術人才。

建造業議會訓練學院所教授的課程包括：

兩年制課程：
    1. 泥水粉飾科
    2. 粗細木工科
    3. 油漆粉飾科
    4. 水喉潔具科
    5. 雲石裝飾科
    6. 金屬工藝科

一年制課程：
    1. 建造棚架科
    2. 機械維修科
    3. 電器裝置科

學徒均可參與實際訓練工作坊和課室授課，且無需交付學費。每名學徒在受訓期間都可取得每月津貼。地盤實習訓練還可以得到特別津貼，以鼓勵更多人從事建造行業。

New frontiers in construction

'Sky City 1000', a 1000 m tall super tower conceived as a vertical composite urban community, was proposed in Japan in 1989. Concave layered structures, referred to as space plateaus and realized as artificial terraces, surrounding a recreational/communal atrium were proposed. A 'conic shell structure' instead of conventional columns and beams was also proposed. Composite panels made of carbon fibre, stronger and lighter than reinforced concrete, were proposed for use in construction. High-tech construction using robots was effective in overcoming labour shortage.

Three-storey elevators (triple-deckers) were proposed to connect each space plateau. Spiral monorails were proposed to circulate between space plateaus.

If completed, 'Sky City 1000' would accommodate 35,000 residents and provide office space for 100,000 people. The total construction period was estimated to be 14 years.

The 'Sky City 1000' concept was proposed in response to high land prices in Tokyo (as high as ¥50 million per m^2) and the overcrowding that was destroying the local environment. Subsequently in 1995, the Japanese proposed a 4 km high megastructure, X-seed 4000, that looked like Mount Fuji. With similar environmental issues, Hong Kong may need to come up with new ideas to tackle future threats that bring both problems and opportunities for innovative developers, architects, engineers, and contractors.

建築的新紀元

「天空之城1000」是日本在1989年所構思的1,000米摩天高向城堡。從畢直的軸線作座標轉出的圓形層面結構，成為每一組的空間平原，並作為中間的休憩/商業中庭，而錐型總結構取代傳統的柱和樑。所運用的合成纖維板，亦比傳統的混凝土更堅固和輕巧。同時，還配合多功能的機械操作，解決勞工短缺的問題。三層的電梯把「空間平原」連接起來，旋轉單軌鐵路亦在空間平原中穿梭往來。據估計此項「天空之城1000」要花十四年的時間興建，建成後，將提供三萬五千個住宅單位，十萬名就業者的辦公室空間。

日本所構想的「天空之城1000」設計，及後來1995年的X-seed 4000，四公里高，形態像富士山的超級結構，目的是解決由於日益昂貴的土地而構成的密集空間，以及因此引致許多的環境污染問題。香港面對同樣的環境問題，亦應該提出創意理念，為未來發展，同時可為發展商、建築師、工程師和承建商提供創新的機會。

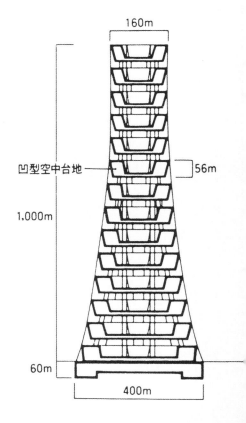

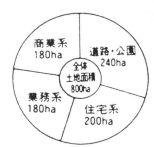

'Sky City 1000': a hypothetical architectural concept in Japan announced in 1989

Chapter 2
Trades of Materials and Technology
工料的類別及技術

Wong Wah Sang 黃華生

2.1 Preliminaries
整體施工預備工作

Construction of a tower

Preliminaries usually form the first section of the specifications, which is part of the tender document and will subsequently become part of the contract document. Items included in the preliminaries are:

1. Form of contract
2. Definition of terms
3. British standards and equivalent standard
4. Manufacturer's recommendations
5. Government regulations
6. General obligations
7. Temporary works and services
8. Administration, insurance, and attendance
9. Materials and workmanship

整體施工預備工作是招標文件中有關訂定規範的第一個章節部份，當協定達成之後，有關方面正式簽署文件和合同，便成為正式的合同書，同時具有法律效力。施工整體預備工作包括：

(1) 合同條款
(2) 名稱的定義
(3) 英國制度指標和國際認可的標準
(4) 製造商的建議
(5) 政府法律條例及標準
(6) 一般須遵守的規範
(7) 臨時工程工作及臨時水電設備
(8) 行政費、保險項目及其他
(9) 物料準則及工藝規格

Contract, standards, and regulations

The Standard Form of *Building Contracts: Agreement and Schedule of Conditions of Buildings Contract for Use in Hong Kong* (with or without quantities), published by the Hong Kong Institute of Architects, Hong Kong Institute of Construction Managers, and the Hong Kong Institute of Surveyors in 1976, is the basis to govern the execution of the construction work for the past four decades. This was updated in 2005 to deal with the new period in the construction industry. The latter version was written in simple English, with shorter paragraphs and subheadings for easy reading. New clauses were added to protect both employers and contractors. In addition, legal terms were more clearly defined for better understanding.

For government projects, a contract called 'General Conditions of Contract for Term Contract for Building Works' is issued by the Development Bureau.

Murray Building, where the Building Authority locates

合同、規格及法律條例

1976年由香港建築師學會、香港營造師學會及香港測量師學會出版的《香港建築合約條例及規格》，是過去數十年作為監管私營發展建築商從事建築工程的基本法律文件。為了迎合建造業的新時代，2005年修訂了新版。新版用簡潔的英文撰寫，輔以短段落和副標題，令條文變得更加易懂。新增的條款適用於同時保護發展商和建造承包商，且定義清晰明瞭。

至於政府工程，一般情況下使用的建造合同，則由發展局發出的文件為依據。

British Standards and Codes of Practice are available from the British Standards Institution (www.bsigroup.com). These include all amendments, revisions, and standards superseding the standards listed. The British standards are available for inspection at the following locations in Hong Kong:

1. Hong Kong Standards and Testing Centre
2. Main Library of the University of Hong Kong
3. Works Branch Library, Murray Building
4. Urban Council Yau Ma Tei Library, Kowloon

Government regulations for construction: Buildings Ordinance (Chapter 123) consolidates regulations such as the Building (Administration) Regulations, the Building (Planning) Regulations, and the Building (Construction) Regulations. Both the authorized person and the registered contractor have a duty to supervise the building works, notify the Building Authority of any contravention of regulations, and comply with the ordinance in general. The difference is that the authorized person will take periodic supervision, but the registered contractor will take continuous supervision of the building works. Once the authorized person has submitted Form BA4 and the registered contractor has submitted Form BA10, the legal obligations are sustained.

Furthermore, as the design and construction of buildings becomes more complex, it is necessary to exercise closer supervision during construction. Section 17 of the Buildings Ordinance provides that conditions may be imposed when approval of plans or consent to commence buildings works is given; qualified supervision is required at certain stages of construction or for some particular operations.

英國制度指標和實際工程規格，則由英國國家標準局制定。當中包括最新的修訂和內容。在香港可供查詢的地方包括：

(1) 香港品質及測試中心
(2) 香港大學圖書館
(3) 美利大廈的拓展署圖書中心
(4) 九龍油麻地市政局圖書館

政府訂定的建築物條例（第 123 章）包括許多有關行政、計劃、施工等法例條文細節。認可人士及註冊承建商需監察建築地盤施工的工程，履行責任知會建築事務監督有關非法的事項、工程。認可人士及註冊承建商在監察工程的責任分別，後者必須在工程期間作全面監察，而認可人士要定期監察該項工程是否遵守一切法律的規格和準則。在預備施工的地盤，認可人士必須呈交表格 BA4，而承建商亦應於開工前呈交表格 BA10，申請作為地盤工程的負責人。

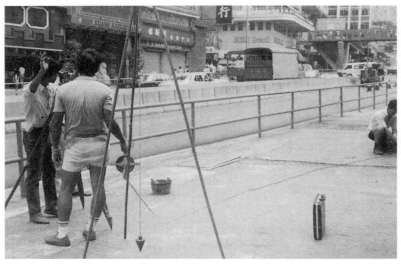

Setting out

General obligations

General obligations include setting out on the site works, compliance with safety regulations as required by the Labour Department and the Industrial Safety Training Centre, maintenance of roads, and protecting or shoring up adjoining properties.

The site boundary can be set out by the Government Survey Office with critical marks identified with marking nails on site. These nails must be properly protected by the contractors after handing over to the landowner from the government. It is advisable to use lending points using triangulation to be marked for easy location of the original setting out points. Levels are often referred to as Principle Datum (PD). Information can be obtained from the nearest government benchmarks. Inaccuracy of setting out can lead to abortive work or delay in progress. Maintenance of roads includes cleaning the wheels and underside of lorries and making sure that they are free of earth before they leave the site, filtering earth, discharging concrete water into a public sewer, and securely stacking rock or debris on lorries to avoid dislodging and falling onto public roads. In some cases, limitation to the loading of vehicular traffic using the public roads has to be observed.

Shoring against adjacent properties or roads is in accordance with the approved shoring plans. Shoring systems such as flying shore, raking shore, and dead shore may be used. Monitoring devices for recording any settlement may also be required. These are all directed for the general safety of building construction.

一般法則規格

一般非列明法則，包括在工作地盤進行量度及定界線，遵守一切由勞工處以及建築業管理局等訂定的規格，地盤附近的公共地面維修、保護及鄰近建築物的臨時支撐等。

地盤的界線由政府測量工程處負責，經測量後再以鐵釘直插於地上作永久測量標誌。承建商需負責保護及以此界點作地盤的範圍。每個定點可用普通的三角幾何學來制定借點及計算實際的方位。地面的平水

Construction site at Laguna City, Kwun Tong

Various hoarding designs, Kingswood Estate in Tin Shui Wai

通常以平面為準則,可以利用附近政府所定的記號量度。準確的定界和量度可避免工程錯誤或延遲。承建商亦需負責地盤附近的地面清潔和安全問題。所有進出地盤的泥頭車,必須清洗妥當,才不易沾滿地面的沙泥;必須每天清除堆積的混凝土和沙石等雜物,避免阻礙公眾通道。支撐隔鄰建築結構的方案,必須經建築事務監督批核。支撐結構方法可分為:(1) 懸空支撐(飛頂);(2) 斜支撐(斜頂);(3) 固牆豎木(死頂)。

施工期間,亦須注意四周的結構是否影響到鄰近的建築物。專門的測試儀器和方法,可以提供定期報告和防止地盤意外。

Temporary works and services

Temporary works and services include hoardings and fencing, scaffolding and signboards, temporary offices and storage sheds, contractor's sheds and guards, and temporary water and electricity supply.

Hoardings and fencing are required to be constructed before the commencement of works. An authorized person is required to submit Form BA 19 and hoarding plans to the Building Authority in order to receive a Hoarding Permit (Form BD 109).

The Hoarding Permit requires the permittee to inform the corresponding electrical company and the PCCW-HKT Telephone Limited (formerly the Hong Kong Telephone Company) 72 hours before commencement of work to ensure that no cable will be affected. An Excavation Permit is also required from the chief engineer, Highways Office, before opening up the footpath or carriageway. Timber or steel structures can be used where appropriate. The project signboards will include English and Chinese, the artist's impressions, or graphics and logos. The contractor must submit Form BA18 together with construction details to apply for a contractor's shed. Contractors prefer containers for temporary offices on larger sites.

Temporary water can be applied and obtained via metered supply from the Water Authority. Temporary electricity is also available from the electric company upon application, the amount of which will depend on the actual necessity.

臨時工作及設施

所有臨時工作及設施包括圍板、搭棚、水牌、臨時建築地盤、物料貯存室、承建商的地盤辦公室、看更房、臨時水電等。

保護地盤的圍街板、有蓋行人道或門架,必須在屋宇署同意展開工程前建造完成。

認可人士必須呈交表格 BA19 和圖紙給屋宇署,以取得工程認可(表格 BD109)。同時亦須向路政署申請掘路紙,才可以在地盤界線之外進行任何挖掘的工作。在建築事務監督發出圍街板的開工紙的同時,更要在 72 小時之前知會香港電燈公司或中華電力公司以及相關電訊公司,確認所有的電源已經截斷。木板和鋼鐵可以用作興建圍街板。通常在當眼的地方會張貼有關該項發展的中英文基本資料,包括承建商、認可人士的公司名稱、其他參與建築計劃的顧問及業主等。

如承建商需建臨時辦事處,必須向建築事務監督入紙申請。在一般大型的建築地盤,承建商多會以貨櫃箱作為臨時工作地方。至於臨時水電可分別向水務署和電力公司申請。

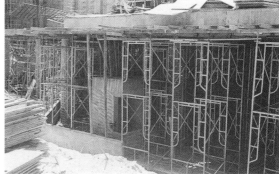

Temporary supports, Kingswood Estate in Tin Shui Wai

Contractors' sheds, Kingswood Estate in Tin Shui Wai

18 Trades of Materials and Technology

Some work areas, Kingswood Estate in Tin Shui Wai

Some construction techniques, Kingswood Estate in Tin Shui Wai

Administration, insurance, and attendance

Regular site meetings are held to monitor the progress of different trades and to coordinate various subcontractors.

Insurance means third-party insurance, employees' compensation insurance, contractor's all-risk insurance, and surety bonds.

Attendance for the nominated subcontractors, nominated suppliers, specialist contractors, government departments, and public utility companies are provided according to the following:

1. Use of plant, ladders, scaffolding, etc. as erected by the contractor. There is no obligation to retain such facilities longer than the contractor's own use.
2. Provision of space and lock-up rooms for storage of materials and tools.
3. Temporary water and electric supply.
4. Guards to safeguard the site.
5. Coordination with the contractor's programme of works.
6. Coordination with other trades such as provision of openings and chases, cuttings.
7. Cleaning and clearing away debris.

行政費、保險項目及其他

定期的地盤會議，能監管不同項目的工作進度以及協調各項目承建商的合作問題。保險項目泛指第三者保險、僱員勞工保險、承建商意外賠償保險及擔保金。

指定的其他項目承建商、物料供應商、特別項目顧問、政府部門、公共設施公司等可共同享有以下的設施，包括：

(1) 地盤的機械、梯級、竹棚等由總承建商負責的項目，但卻絕不能超過承建商工程訂立所運用的時間；
(2) 由承建商提供的貯物空間；
(3) 臨時水電的設備；
(4) 地盤看更；
(5) 配合承建商的施工計劃；
(6) 配合其他工作項目的施工過程；
(7) 清潔及移除一切廢料。

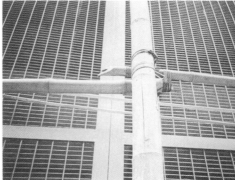

Traditional bamboo scaffolding

Materials and workmanship

In simple terms, the quality of work is required to be consistent with good building practice in Hong Kong and to comply with the relevant British Standards (BS) or Codes of Practices (CP) unless otherwise specified. Sometimes it is written in the specifications that quality must in every respect be to the satisfaction of the architect, giving the architect power to reject any 'unsatisfactory' work.

Submissions of samples, execution of mock-up, and preparation of shop drawings are all good practice for understanding between all parties before actual construction work begins.

Cost of preliminaries

Some of the preliminaries items can usually be priced between 5% and 10% of the total contract sum.

An example is quoted for the breakdown of different trades of a cost estimate for an office building and is tabulated below:

物料和施工質素

概括地說,優質的工作是必須符合香港的建築規範,如英國的國際性標準或其他同等的指標。有時亦可以把要求的準則詳細列明,但要符合建築師的要求和指示。在合同條文中已列明建築師有權批核承建商的工作質素。

所謂優質的工作,包括了物料的選擇、提供施工圖讓建築師批閱,以及其他項目承包商或顧問的合作等問題細節。

施工整體預備工作費用

有些施工整體預備工作項目在初步預算表中,可以明確地訂定費用和價錢。預備工作費用通常佔總建築費約百分之五至十左右。以下是一座商廈建築費用的明細計算表:

Proportioning of cost estimated for an office building of about 12,000 m² floor area
初步估算一座約12,000平方米商廈建築費用明細

Section (項目)	% Cost (費用百分比)
1. Foundation 地基工程	2.8%
2. Substructure 地面以下的結構工程	1.3%
3. Superstructure 地面以上的結構工程	11.9%
4. Facade finishes 大廈外牆的物料	25.8%
5. Plumbing and drainage 渠務工程	1.6%
6. Architectural works 建築工程	16.5%
7. External works and landscaping 園林及周邊工程	0.4%
8. Preliminaries 前期施工	6.5%
9. Electrical 電器	5.1%
10. Air-conditioning 空調設施	13.9%
11. Fire services 消防設施	3.2%
12. Lifts and escalators 電梯和電動升降電梯	6.9%
13. Contingencies 預備費用	4.1%
Total 合共	100%

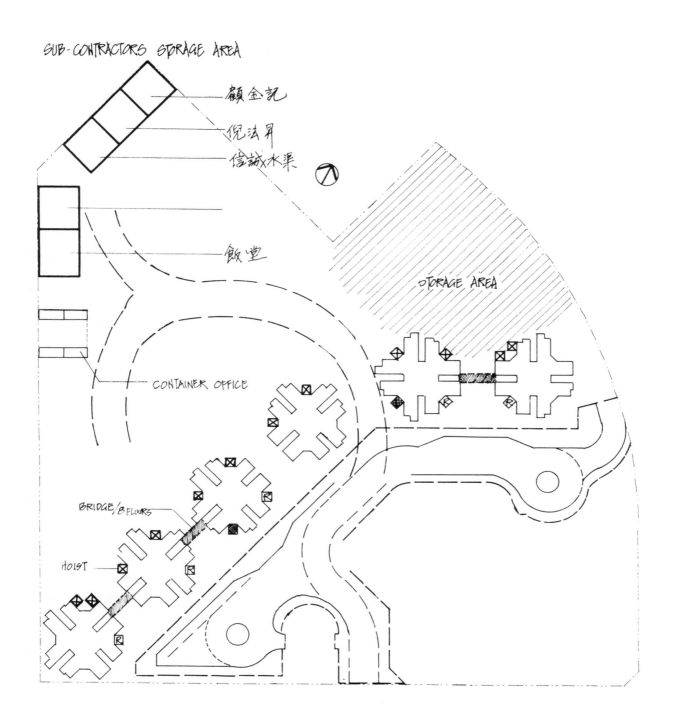

Layout plan showing constractor's site organization, Kingswood Estate at Tin Shui Wai

2.2
Demolition and Excavation Works
拆樓及挖掘工程

Commencement of demolition for some old tenement houses. Erection of screens and scaffolding.

Heavy machines imposing great loads on old buildings may not be permitted for demolition.

Preparation work

Before the commencement of demolition, a thorough survey is done and available drawings of structures are examined.

The authorized architect will arrange for disconnection of utilities services including gas, electricity, water, and telephone. Any fitting attached to the building in connection with tramway services, street lighting, electricity or other services is removed. Adjacent road signs, parking meters, hawker stalls, street lighting, etc., affected by the demolition are removed by the relevant government departments.

Apart from the hoarding, the contractor will erect at walls abutting any street, fans, or catch platforms at the first floor level or at levels as that may be necessary to prevent materials from falling to the gound. Dust screens are required to cover the whole wall area. Sewer and drainage connections are sealed, and glazed sashes and doors are removed before breaking down the structures. If necessary, shoring may be erected to prevent the accidental collapse of any part of the building or any adjoining building.

預備工作

樓宇拆卸工程進行前，應先檢查該樓宇及鄰近樓宇的結構和安全等問題。

政府認可的專業人士要負責有關拆樓前所需要的準備工作，包括通知有關部門暫停供應公共設施，如水、電、煤氣、電話等外，還要留意其他有關的設施，如電車服務、路燈、道路指示牌、泊車位、小販檔位等。

除了圍街板外，承建商需要在地盤向街的地方，由一樓起興建承接平台，以防範拆樓時跌落的沙石和雜物。此外，要在被拆卸樓宇的周圍架起防塵網。而有關渠務、水管等當然都要完全封閉妥當。至於門、窗及玻璃，都必須先拆除才可以進行其他的拆卸工程。

有時，還需要安裝支撐架構來保護鄰近的建築物，以避免因拆樓工程而導致鄰近樓宇倒塌。

Demolition works

The demolition works must be carried out under the supervision of an experienced person and the required procedures as stated in APP21. A notice in Form BA20 stating the name of the person in charge must be posted in a prominent position on the demolition site. Competent workers are employed for the demolition of the structural frame, roof, staircase of a building, areas with risk of collapse, and for the cutting-up of structures.

It is essential that debris or materials not accumulate on floors so that floors are overloaded, as floors are not designed for such heavy loads. Great pressure exerted on adjoining buildings and on hoardings is not desirable either.

Some demolition methods, such as the use of a crane and a hammer, are considered to pose unacceptable risks or impose heavy loads on the building and are therefore unsuitable in certain circumstances.

In the demolition of buildings containing asbestos-based products, special care is exercised. The area concerned is cut off by dust-proof screens, and all workers must wear half-mask dust respirators.

Timber raking shore

拆樓工程

拆樓工程必須在經驗豐富的工程人員監管下進行，同時要履行《認可人士及註冊結構工程師作業備考》APP21 條內所說明的一切工作形式。尤其是當要拆除樓層的結構架、樓梯，或者是該部份有倒塌的危險，又或是涉及結構等的項目，都要聘請有經驗的技工。因此，在每個拆樓工程的地盤，都需要在當眼的地方張貼屋宇署表格 BA20，清楚列明地盤監管人的姓名資料。

當每層的拆卸物堆積而構成層面的承重力增加，或使圍街板和鄰近的建築物構成沉重的壓力，都是很危險的，因此，要定時把現場的堆積物移離。

有些拆樓方法，需運用到起重機、鐵槌等，會被列為危險類別。除非在不得已的情況下，否則不應考慮。

如拆樓工作涉及含「石棉」成份的建築物料，更需加倍謹慎小心。勞工處及環境保護署會要求承建商提交拆卸石棉工程的方法及處理方式，而該地方更要以防塵帳幕四面包攏起來，所有的工人則要配戴氧氣面罩。

Timber flying shore

Excavation

Excavation includes removal of the existing old foundation, cutting off pile heads to the required level, and removing and legally disposing of all surplus materials.

Planking, strutting, or shoring may be required to support the sides of the excavation. Work is carried out to prevent erosion or slips. Working faces are limited to safe slopes and height. Material must not be stockpiled to cause a landslide, and authorized persons, registered structural engineers, are required to take adequate precautions to ensure public safety whenever necessary during excavation as part of the work as stated in APP57.

It is essential that earthwork be sealed up after each day's work or when heavy rain is expected. Cutting is planned to prevent ponding. Temporary water courses or channels are provided to discharge the water, usually through a sand trap before going into permanent drains.

挖掘工程

挖掘工程包括：

(1) 清除原有的地基；
(2) 切斷樁頭到指定水平；
(3) 清除所有剩餘的雜物。

在挖掘地方的兩旁，需要以木板或其他支撐方法來預防兩旁的結構倒塌。工作的範圍受安全高度和斜面所限制，挖掘的雜物不能堆積於任何地方而引致山泥傾瀉。而各認可人士及有關工程師應根據《認可人士及註冊結構工程師作業備考》APP57 所規定，作出適當的設計以確保公眾安全。

每天挖掘工程完成後，要小心把挖掘的地方覆蓋妥當。加設臨時水渠可疏導因工程或滂沱大雨後所引致的積水，而每個疏水導口都要裝置隔沙井，才可以接駁公共渠道。

Steel sheet piling for basement excavation

Timber planking and strutting for trench excavation

Steel strutting for basement excavation

Dewatering

In constructing foundations and basement excavations, dewatering may be required to achieve a dry, workable base, during which precautionary measures are taken to avoid impairing the stability and causing undue settlement of any adjoining buildings, streets, and land. Ground treatment such as grouting may be necessary.

In connection with dewatering, adjoining foundation details have to be studied together with site investigation report. All the related information, plans, reports, and assessments should submit to the Building Department. An assessment of dewatering effects is monitored by piezometric and settlement records. Limiting criteria for movements and groundwater pressures are set up. If these values are reached, action such as shoring or underpinning may have to be taken.

抽水工程

如在地下層或地基進行挖掘工程，都需要抽取地下水，以使地盤乾爽，方便工作。不過，抽水工作要有適當的保護，以免影響鄰近水位以及建築物的地基。一般地面的處理方法如灌漿等，都可以使鄰近水位保持不變。

設計抽水工程計劃時，須要一併研究鄰近地基結構和探土報告等資料，並應向建築署提交所有相關研究的資料、記錄詳情及評估。抽水工程進行時，要定期查閱水壓計與沉陷觀察釘，觀察工程對附近地區所帶來的影響並作報告。若測度結果超過計算所定的指標，必須採取適當的應變方法，如附加支撐托等。

Earth work machines

26   Trades of Materials and Technology

Retaining wall work: chunam plaster

Rock cutting with concrete guniting surface

Pipe pile wall

Caisson wall

2.3
Concrete Work
混凝土工作

Concrete is composed of aggregate mixed with cement and water. The proportion of each material controls the strength and quality of the resultant concrete. Steel reinforcement is added to acquire the strength needed to stand tension.

混凝土由細沙、水泥和水合成。成份比例不同會混合出不同強度和品質的混凝土。鋼筋附加在混凝土內以加強對拉力的承受。

Cement

Cement is the setting agent of concrete and can be ordinary Portland cement, rapid-hardening Portland cement, or Portland blast-furnace cement. These kinds of cement are made up of chalk or limestone and clay. High-alumina cement, which hardens very fast, is not used in general structural concrete works. Other types of cement require approval by the Building Authority before use. The setting time for cement is not less than 45 minutes initially and not more than 10 hours at the final stage. The architect may request the contractor to submit certificates of origin and quality tests for approval. The source of cement must not be changed. On site, cement is stored in a dry, weatherproof store with a raised floor.

水泥

水泥是混凝土的凝結劑。水泥可分為普通卜德蘭水泥、特別卜德蘭水泥和高爐水泥。它們主要由白亞、石灰砂漿或黏土組成。水泥中鋁金屬成份越高，混凝土的凝結速度會越快，因此含鋁質較高的水泥，都不會用來作一般的結構混凝土。除上述所提到的水泥，其他類別的水泥都需要建築署的批准。水泥凝結的時間，開始時至少要 45 分鐘，但卻不能超過 10 個小時。建築師可能會要求承建商提交水泥的產地來源證明及品質保證等。地盤裏的水泥要貯存在乾爽和不受雨打風吹的地方，有時須用架空地台安放。

Water

Water for concreting must be clean and uncontaminated potable water from the government main supply. The amount of water used for making concrete should not exceed the specified criteria. The water-cement ratio in weight is usually between 0.4 and 0.7.

Water is tested by the initial setting time test and a compressive strength test. For the initial setting time test, a test block is made with water. The appropriate test cement and the initial setting times of the cement should not differ by more than 30 minutes.

For the compressive strength test, concrete test cubes are made with water. The appropriate test cement and the average compressive strength of the concrete test cubes must not be less than 90% of the average strength of the control test cubes.

Certain impurities in water are permissible. Chlorides must not exceed 50 mg per litre. Sulphates must not exceed 1,000 mg of sulphur trioxide per litre. Alkali carbonates and bicarbonates shall not exceed 1,000 mg per litre. Sea water if used in cement will cause surface dampness, efflorescence, and a moderate reduction in strength.

水

專為混凝土而用的水,必須是清潔及不含雜質的飲用水。水的份量不能超過指定的標準。一般而言,水與水泥重量的比例是由 0.4 至 0.7。

清水的檢定方法分為第一次凝結時間測試和壓力測試。凝結測試是把水與適當份量的水泥混合,得出的凝結時間不能超過 30 分鐘。

至於壓力測試的標準,是用某份量的水泥和水製成的模擬混凝土磚的平均強度,應不得少於指定的標準數的百分之九十。

水中含雜質的標準為每公升:

氯化物不得超過 50 mg;

硫化物中的硫化氧不得超過 1,000 mg;

鹼性碳酸鹽及碳酸鹽不得超過 1,000 mg;

一般而言,海水不能用於混合水泥作建築的用途,因為海水中的物質可引致混凝土的表面受潮,造成風化現象而起白花,同時減低了混凝土的堅固程度。

Concrete batching plant

Compressive strength of cements (BS 12:1989)
水泥的受壓強度表

Cement type 水泥的類別	Test age 測試時間 (days) (日)	Strength	
		Concrete 混凝土 (N/mm^2)	Mortar (N/mm^2)
Controlled fineness Portland 特幼細卜德蘭水泥	3 (72 hours)	Not less than 33 不少於33	Not less than 23 不少於23
	28	Not less than 29 不少於29	Not less than 41 不少於41
Ordinary Portland 卜德蘭水泥	3 (72 hours)	Not less than 15 不少於15	Not less than 25 不少於25
	28	Not less than 34 and more than 52 介乎34與52之間	Not less than 47 and not more than 67 介乎47與67之間
Rapid-hardening Portland 特強卜德蘭水泥	2 (48 hours)	Not less than 15 不少於15	Not less than 25 不少於25
	28	Not less than 28 不少於28	Not less than 52 不少於52

Aggregate

For plain concrete, aggregate consists of sand, well-burnt brick, well-burnt tile, well-burnt clinker, stone, or other approved materials by the Building Authority.

For reinforced concrete, aggregate must be hard, strong, clean sand, and crushed stone. As fine aggregate, at least 90% by mass will pass through a 5 mm mesh screen, and not more than 15% by mass will pass through a 150 um mesh. As coarse aggregate, not more than 10% by mass will pass through a 5 mm mesh screen, and at least 95% by mass will pass through mesh of a size 6 mm less than the minimum lateral distance between reinforcing bars, or 6 mm less than the minimum cover, whichever is the smaller. In case of solid slabs, the standard is reached when those aggregates can pass through a 20 mm mesh screen.

粒 料

純混凝土的粒料由沙、燒過的磚塊和瓦片、熔碴、石或其他經建築事務監督批准的用料所組成。

結構用的鋼筋混凝土中的粒料必須是堅硬、清潔的沙料或碎石。比較細微的粒料，需含超過百分之九十的質量可穿過 5 毫米濾網，及不超過百分之十五的質量可穿過 150 微米特微細網。但較粗的粒料的定義則只要少於百分之十的質量可穿過 5 毫米濾網，及至少百分之九十五的質量可穿過少於兩條補強鋼的橫面距離 6 毫米，或者是 6 毫米少於混凝土的最少保護層，而以二者最少的數位為準。在實心的樓板結構裏，標準為通過 20 毫米孔洞的網。

Test cement (BS 3148:1980)
水泥測試

Type of cement 水泥的類別	British Standard number 英制標準度	Minimum initial setting time 最少的初步凝結時間	
		In the BS 英制標準 minutes (分鐘)	For this test 這測試 minutes (分鐘)
Ordinary Portland cement 普通卜德蘭水泥	12	45	75
Rapid-hardening Portland cement 快硬固卜德蘭水泥	12	45	75
Portland blast-furnace cement 爐燒卜德蘭水泥	146	45	75
Sulphate-resisting Portland cement 防硫化物的卜德蘭水泥	4027	45	75
Supersulphated cement 超硫化水泥	4248	45	75
Low-heat Portland cement 低熱能卜德蘭水泥	1370	60	90
Low-heat Portland, blast-furnace cement 低熱能爐燒卜德蘭水泥	4246	60	90
High-alumina cement 高鋁質水泥	915	120	150

Stockpile of aggregate

Crushing and sorting out aggregate

List of tests for aggregate by British Standards Institution
英國國際標準的粒料測試表

BS 812	Part 1	Methods for sampling and testing of mineral aggregates, sands, and fillers: sampling, size, shape, and classification	
	第1部	礦物粒、石、填料的抽樣和測試方法—抽樣、度數、形狀、分類	
BS 812	Part 2	Methods for sampling and testing of mineral aggregates, sands, and fillers: physical properties	
	第2部	礦物粒、石、填料的抽樣和測試方法—物理性	
BS 812	Part 3	Methods for sampling and testing of mineral aggregates, sands, and fillers: mechanical properties	
	第3部	礦物粒、石、填料的抽樣和測試方法—機械性	
BS 812	Part 4	Methods for sampling and testing of mineral aggregates, sands, and fillers: chemical properties	
	第4部	礦物粒、石、填料的抽樣和測試方法—化學性	
BS 812	Part 100	Testing aggregates: general requirements for apparatus and calibration	
	第100部	測試部份—工具與度尺基本要求	
BS 812	Part 101	Testing aggregates: guide to sampling and testing aggregates	
	第101部	測試部份—如何抽樣與進行測試	
BS 812	Part 106	Testing aggregates: method for determination of shell content in coarse aggregate	
	第106部	測試部份—粗粒料中的貝類物質等	
BS 812	Part 109	Testing aggregates: method for determination of moisture content	
	第109部	測試部份—水份含量的定法	
BS 812	Part 110	Testing aggregates: method for determination of aggregate crushing value (ACV)	
	第110部	測試部份—定碎石的量度	
BS 812	Part 112	Testing aggregates: method for determination of aggregate impact value (AIV)	
	第112部	測試部份—測試受壓程度的方法	
BS 812	Part 113	Testing aggregates: method for determination of 10 per cent fines value (TFV)	
	第113部	測試部份—測試應變程度的方法	
BS 812	Part 114	Testing aggregates: method for determination of the polished stone value	
	第114部	測試粒料部份—測試面石量度的方法	
BS 812	Part 117	Testing aggregates: method for determination of water soluble chloride salts	
	第117部	測試粒料部份—定水溶氧化鹽的量度的方法	
BS 812	Part 118	Testing aggregates: method for determination of sulphate content	
	第118部	測試粒料部份—定硫化物的方法	
BS 812	Part 119	Testing aggregates: method for determination of acid soluble material in fine aggregate	
	第119部	測試粒料部份—定酸溶解性的微粒量度法	
BS 812	Part 120	Testing aggregates: method for testing and classifying drying shrinkage of aggregates in concrete	
	第120部	測試粒料部份—粒料在混凝土中的乾燥收縮速度的測驗和方法	
BS 812	Part 121	Testing aggregates: method for determination of soundness	
	第121部	測試粒料部份—定優劣的分類法	
BS 812	Part 124	Testing aggregates: method for determination off rost heave	
	第124部	測試粒料部份—定霜、隆距的方法	
BS 812	Part 103.2	Testing aggregates: sedimentation test	
	第103.2部	測試粒料部份—沉澱測驗	
BS 812	Part 105.1	Testing aggregates: flakiness index	
	第105.1部	測試粒料部份—鱗片的指標	
BS 812	Part 105.2	Testing aggregates: elongation index coarse aggregate	
	第105.2部	測試粒料部份—粗粒的可伸展程度	

Drum mixer for site concrete mix, e.g., caisson rings

Pumped concrete being placed on site

Concrete mixes

Concrete mixes are expressed as proportions of cement, fine aggregate, and coarse aggregate, e.g., 1-2-4 nominal mix = 1 part cement, 2 parts fine aggregate, and 4 parts coarse aggregate.

The Building (Construction) Regulations have concrete of designed mix and prescribed mix. The latter is used for minor structural and non-structural works.

混合混凝土

混凝土成份的說明通常都是按著水泥、微粒料及粗粒的成份來標明的。例如：1-2-4 的通常混合數，為一份的水泥中用兩份的微粒及四份的粗粒。

建築條例中的「建築工程」章節清楚規定混凝土的混合質包括：「配料經設計」及「訂明配料」。後者應用於小型結構及非結構工程。

Mixing and placing

Mixing of concrete is normally carried out in a power-driven batch-type mixer. The batch capacity, method of loading, time of mixing, and speed of operation are set as recommended by the mixer manufacturer. However, a minimum of two minutes is required for mixing.

Ready-mixed concrete batched off site is carried in purpose-made agitators operating continuously, or in truck mixers. The concrete is placed within 2.5 hours of the introduction of cement to the aggregates and within 30 minutes after discharge from the agitator or truck mixer.

Generally, concrete is deposited before setting has commenced and without segregation of the materials. There should be adequate consolidation by tamping and vibrating, after which the concrete should remain undisturbed and adequately protected from the weather. In hot weather steps are taken, including sheltering, lowering the temperature of mixing water and aggregates, or using cold cement, to maintain the mixed concrete below 35oC. Curing of concrete is usually done by thorough wetting and covering with a layer of approved waterproof paper or plastic membrane for a minimum of four days.

混合與澆灌混凝土

混合混凝土的工作由電動混合機分批進行。混合機的容量、工程方法、混合時間、速度須由供應商負責訂定，但混合時間絕不能少於二分鐘。

預備混合的混凝土，在特別設計的攪拌機中不停地轉動攪拌。混合混凝土的程序和時間是，先將水泥與粒料混合攪拌約兩個半小時，而在三十分鐘之內得將全部混合成的混凝土澆灌。

一般而言，混凝土會在凝結前沉澱，而不會把不同的物料分解回原來的模樣。因此，在澆灌期間，要保持著震盪和搗固，才可以讓它自然的凝固，而且要避免日曬雨淋。在炎熱的天氣下，混凝土須收藏在比較陰涼的地方；有時要用冷卻的水或水泥來保持它的混合溫度至 35 ℃ 以下。混凝土熟成的最後步驟，是先切底澆灌去保存濕度，再以合規格的防水紙或塑膠布覆蓋四天以上。

Reinforcement

Reinforcement for concrete is specified in conformity with the appropriate British Standards. Hot rolled steel bars, hard drawn mild steel wire, cold worked steel bars, and hard drawn steel wire fabric can be used. The contractor supplying the reinforcement is required to submit certificates of origin and chemical analysis from the steel manufacturer. Routine test results for each consignment of steel reinforcement supplied to the site should be made. Test specimens can be sent to the Hong Kong Laboratory Accreditation Scheme (HOKLAS) accredited for testing.

It is important that test reports verify the mass and tensile properties of the steel. They must be submitted to the Building Authority within 60 days of the delivery of the steel reinforcement to the site.

Steel reinforcement must be free from loose mill scale, loose rust, oil grease, or other matter likely to adversely affect the bond with concrete. Before placing in position, the steel is cleaned by wire-brushing.

鋼 筋

混凝土中的鋼筋是根據英國標準來分類，主要分為熱轉壓鋼筋、硬拉普通鋼線、冷作鋼筋及硬拉鋼筋纖維等。

承建商需要呈報有關鋼筋來源地的證明文件、化學報告等資料作批核及參考記錄。一般鋼筋工程中，供應商要作定期的抽樣測試，以保證物料的質素。在香港，鋼筋可送交香港實驗所許可計劃作指定檢測測試。

測試報告中會列明鋼筋的質量和拉力的特性。報告須於鋼筋在地盤使用之 60 天內呈交建築事務監督備案。

鋼筋須保持清潔，沒有鐵銹漬、油漬，或其他可能影響混凝土強度的物質。因此，在澆灌混凝土前，地盤工人須以清水清洗附在鋼筋上的雜質。

Chemical composition of steel grades
鋼筋的分類

Cast analysis (form BS 1449:1988)
鋼筋的分類：(表BS1449:1988)

Element 成份	Grade 250 250級	Grade 460 460級
	% max. 上限百分比	% max. 上限百分比
Carbon 碳	0.25	0.25
Sulphur 硫	0.060	0.050
Phosphorus 磷	0.060	0.050
Nitrogen 氮	0.012	0.012

Preferred nominal sizes (form BS 1449:1988)
建議之標稱尺寸 (表BS1449:1988)

Grade 級數	Nominal sizes 標稱尺寸
	mm
250	8, 10, 12, 16
460	8, 10, 12, 16, 20, 25, 32, 40

Concrete cover

The minimum thickness of concrete cover to reinforcement is governed by the structural use of the reinforced concrete and the fire-resisting period (FRP) required by the element of construction, which in turn is related to the use and volume of the building. For the latter case, a 15 mm concrete cover can give an FRP up to two hours, while a minimum 25 mm concrete cover is provided for four hours' FRP.

For a general indication, the minimum thickness of concrete cover for reinforcement is tabulated on page 35.

Steel chairs of plain surface and round section can be used to support top reinforcement in slab, rafts, and vertical wall reinforcement. Cover spacers of concrete blocks of similar quality to concrete are also used to support reinforcement in position.

混凝土的保護層

混凝土中的鋼筋距離混凝土表面與其結構應用和物料結構等的防火時間，都是因應不同建築物的應用和空間所決定的。以後者為例，15 毫米厚混凝土保護層的鋼筋組合，可以達到兩小時的防火時間；而 25 毫米的組合，則可以達到四小時的防火時間。

無花紋圓形切面的鋼鐵座可以用來支撐樓板上層的鋼筋和垂直牆中的鋼筋。跟混凝土具有同等性質的混凝土小磚塊亦可用作固定鋼筋保護層厚度。

Precast fibreglass formwork for waffle slab

Steel workers' yard

Minimum thickness of concrete cover for reinforcement
鋼筋給混凝土覆蓋之最小厚度

Specified grade strength 混凝土強度	Reinforced concrete 鋼筋混凝土 Conditions of exposure 外露情況		Prestressed concrete 預應力混凝土 Conditions of exposure 外露情況		Slabs and walls in enclosed buildings 內牆及樓板
	Moderate 受遮蓋 mm	Severe 全露 mm	Moderate 受遮蓋 mm	Severe 全露 mm	mm
20	30	–	–	–	20
25	30	40	–	–	20
30	30	35	30	35	20
35	25	30	25	30	15
40	25	30	20	25	15
45	25	30	20	25	15

Timber formwork

BS 5975 states coordination and supervision of formwork as '[w]ork on site should be the subject of careful direction, supervision and inspection to ensure that the falsework structure is constructed safely in accordance with the agreed design with materials of agreed quality, and that only when all checks have proved satisfactory is the structure first loaded, and then dismantled in accordance with an agreed procedure.'

Timber formwork of solid timber planks or plywood forms the mould against which the concrete is cast, thus acquiring the shape, form, and surface texture needed. Formwork must be strong enough to stand the total weight of formwork, reinforcement and wet concrete, construction loads, construction traffic, as well as wind loads. The formwork itself is usually supported by propping of timber posts or steel struts, which can support greater heights. Though the contractor is responsible for the design and construction of all temporary formwork and supports, the architect and engineer will also make periodic inspections to ensure the formwork is properly erected. Also, the forms should be adequately watertight to prevent loss of material during concrete placing and should be thoroughly cleaned of wood chips and rubbish before concreting.

Formwork for casting circular columns

木工模板工作

英國標準 BS5975 列明有關聯絡和監察模板工作為:「地盤工作都需要有小心的指示、監察、檢查,確保結構工作附合安全標準,遵照認可的物料設計、品質規範,以及符合結構的要求;而一切拆卸工序都依照認可的方法和程序。」

木工的模板工作由實木板或夾板組成,可以用作澆灌混凝土的模板,更可配合立體形狀、表面質感等需要。模板必須堅硬才可以承托模殼、鋼筋、濕混凝土、建築操作的基本重量等等。模板本身由木柱支撐,若用鋼鐵柱更可使承擔的高度增加。雖然承建商負責有關模板設計、施工方式、物料運用,但建築師與土木工程師需定時到地盤監察工人的工作是否符合安全標準。此外,模板必須防漏以避免澆灌混凝土時造成物料的損失,並須清理表面的木屑和塵埃後才澆灌混凝土。

Mixer truck and pump truck

Fixing of steel reinforcement on timber formwork

The erection of metal formwork

Minimum periods for retaining formwork (Portland cement) before striking

As a general guide, minimum periods for retaining formwork (Portland cement) before striking are set out as follows:
一般常用的卜德蘭水泥的木模需維持時間

Vertical formwork to beam sides, walls, and columns (unloaded) 垂直於樑邊、牆和非承重的柱子	3 days 3日
Soffit formwork to slabs (props left under) 層面的底部（維持臨時支柱）	4 days 4日
Props to slabs (unloaded) 層面的臨時柱（非承受重的）	7days 7日
Soffit formwork to beams (prop left under) 樑的底面（維持臨時支柱）	7days 7日
Props to beams (unloaded) 樑的臨時支柱（非承受重的）	16 days 16日
Props to cantilevers (unloaded) 懸臂的臨時支柱（非承受重的）	28 days 28日

Metal formwork

Metal formwork functions similar to timber formwork and is supported by metal proppings. It is made up of steel or aluminium and is commonly reused for tall tower castings, such as for floor slabs and for façade modules. As metal has a longer lifespan than does timber, the metal formwork can be constructed on a greater scale. The installation of the metal formwork is simpler and more systematic. It shortens a construction cycle, gives better quality, and has simpler procedures.

However, metal formwork has its own disadvantages. As metal plates have a smooth surface and less friction, the decorative façade element is not easily formed and attached to the surface after the concrete is cured; therefore, a simple shape and form is recommended.

Also, it is difficult to modify the formwork during construction, so the design of the building has to be confirmed in the early stage. Therefore, this kind of formwork will be most efficient and economical if used by identical towers and/or simple façade buildings with plainer fins.

金屬模板工作

金屬模板工作與木工模板工作相似，分別只是金屬模板由金屬物料製造而成。其中鋁或鋼的材質可以重複使用，因此常用作高樓大廈鑄造，例如樓板和外立面模塊。由於金屬的跨度大於木材，金屬模板可用更大的比例製造。金屬模板的安裝更加簡單和系統化，使得工程週期更短、更簡單且質素更高。然而，金屬模板也有自身的缺點。因為金屬模板的表面光滑，摩擦力小，所以在混凝土固化後，如需於混凝土表層增加裝飾，會比較困難。於是，建築的設計必須在早期階段完成。這種模板對於高樓，或簡單且少裝飾的外牆，會是最高效和經濟的選擇。

Slump test

A slump test aims to determine the consistency of fresh concrete. A galvanized iron mould in the form of the frustum of a cone with a bottom diameter 200 mm, top diameter 100 mm, and height 300 mm is used. The mould is filled with concrete in four approximately equal layers. Each layer is tamped with 25 strokes of a tamping rod (16 mm diameter, 600 mm long with round end for tamping). After the mould is filled, the concrete is struck off and the top finished with a trowel. The mould is removed by raising it slowly and vertically. The slump is measured by determining the difference between the mould height and the highest point of the specimen being tested.

坍落度測試

坍落度測試用於評核剛製成的混凝土的流動性。以一個鉛水鐵模的截角錐體——底部直徑200毫米、頂部直徑100毫米、高300毫米，注滿四層相若的混凝土份量。每一份量的混凝土以搗固棒在外壁向內搗約二十五次（每次的直徑為16毫米，長為600毫米）。當模殼被注滿之後，便取下混凝土的凝固體，另於頂部用泥鏟修平，跟著把鐵模小心提高移開。混凝土的坍落度便是取自模殼的高度與所測試樣品高度的差異。

Concrete tests

Concrete must satisfy the stated resistance to compression at 28 days after mixing. The resistance to crushing at seven days after mixing is used as a control. The site sampling of concrete and the making and curing of standard 150 mm tests cubes must be supervised by an experienced and competent person.

The tests must be carried out by a recognized laboratory and the test reports accompanied with a statement signed by the authorized person or registered structural engineer to confirm that the acceptance criteria have been complied with. This must be submitted to the Building Authority within 21 days after testing.

Records of test cubes are taken to account for the following information:

1. identification mark of each cube
2. designation of concrete
3. brand and type of cement
4. types of aggregate
5. mix proportions
6. water-cement ratio
7. dates of moulding and testing
8. location after placing in the work of the concrete from which the sample was taken
9. number of cubes and their marks
10. mixer from which the sample was taken
11. weather conditions
12. curing conditions

The Building Authority will also conduct tests on site to ensure the resistance strength of completed concrete structures, e.g., at columns, core wall areas are met. This is done by polishing an approximate 150 × 150 mm surface of concrete. A hammer test is carried out to ascertain the required strength.

If the concrete test cubes fail to meet the standard of acceptance, 100 mm or 150 mm diameter core samples are drilled and taken from the finished concrete work for the crushing test. In this respect, cores will be taken for concrete when they are at least 28 days old.

Further failure will lead to the replacement of site concrete thus condemned, and the standard of quality control will be improved.

混凝土的測試

混合 28 天之後的混凝土，必須能承受抗壓測試。壓碎阻力的測試，是以混合之後的 7 天為一個控制指標。在地盤裏，抽取混凝土以及製作標準 150 毫米磚仔測試的報告書，都必須由有經驗的人員負責。

如磚仔測試不及格或懷疑混凝土有問題，可以抽取 100 或 150 毫米直徑芯樣作測試。如再失敗，就需要重造有問題的混凝土。

Common defects

Shrinkage cracks, voids, and honeycomb are among the common defects in concrete causing water leakage and even structural failure in worse situations. The reasons for failure are complicated and may involve any default in the procedure from mixing to placing and curing the concrete. Even early removal of formwork may lead to defects.

Remedial measures, which include cement grouting, epoxy grouting, or guniting, should be recommended and supervised by the registered structural engineer.

Concrete test cubes

Slump test

List of British Standards for testing concrete
英國測試混凝土標準

BS 1881:	Part 1	Methods of sampling fresh concrete
BS 1881:	Part 2	Methods of testing fresh concrete
BS 1881:	Part 3	Methods of making and curing test specimens
BS 1881:	Part 4	Methods of testing concrete for strength
BS 1881:	Part 5	Methods of testing hardened concrete for other than strength
BS 1881:	Part 6	Analysis of hardened concrete
BS 1881:	Part 101	Methods of sampling fresh concrete on site
BS 1881:	Part 102	Methods for determination of slump
BS 1881:	Part 103	Methods for determination of compacting factor
BS 1881:	Part 104	Methods for determination of vebe time
BS 1881:	Part 105	Methods for determination of flow
BS 1881:	Part 106	Methods for determination of air content of fresh concrete
BS 1881:	Part 107	Methods for determination of density of compacted fresh concrete
BS 1881:	Part 108	Methods for making test cubes from fresh concrete
BS 1881:	Part 109	Methods for making test beams from fresh concrete
BS 1881:	Part 110	Methods for making test cylinders from fresh concrete
BS 1881:	Part 111	Methods of normal curing of test specimens (20°C method)
BS 1881:	Part 112	Methods of accelerated curing of test cubes
BS 1881:	Part 113	Methods for making and curing no-fines test cubes
BS 1881:	Part 114	Methods for determination density of hardened concrete
BS 1881:	Part 115	Specification for compression testing machines for concrete
BS 1881:	Part 116	Methods for determination of compressive strength of concrete cubes
BS 1881:	Part 117	Methods for determination of tensile splitting strength
BS 1881:	Part 118	Methods for determination of flexural strength
BS 1881:	Part 119	Methods for determination of compressive strength using portions of reams broken in flexure (equivalent cube method)
BS 1881:	Part 120	Methods for determination of the compressive strength of concrete cores
BS 1881:	Part 121	Methods for determination of static modulus of elasticity in compression
BS 1881:	Part 122	Methods for determination of water absorption
BS 1881:	Part 124	Methods for analysis of hardened concrete
BS 1881:	Part 125	Methods for mixing and sampling fresh concrete in the laboratory
BS 1881:	Part 127	Methods of verifying the performance of a concrete cube compression
BS 1881:	Part 201	Guide to the use of non-destructive methods of test for hardened concrete
BS 1881:	Part 202	Recommendations for surface hardness testing by rebound hammer
BS 1881:	Part 203	Recommendations for measurement of ultrasonic pluses in concrete
BS 1881:	Part 204	Recommendations on the use of electromagnetic covermeters
BS 1881:	Part 205	Recommendations for radiography of concrete
BS 1881:	Part 206	Recommendations for determination of strain in concrete
BS 1881:	Part 209	Recommendations for the measurement of dynamic modulus of elasticity

Voids and honeycomb

一般錯誤

混凝土中常見的錯漏包括收縮的裂縫、空洞「蜂巢」等，都會引致混凝土漏水和影響結構的能力。有許多原因引致這些情況，比如在混合混凝土的過程中出錯或在凝固時出現了問題，甚至在拆移木板時疏忽等。

補救的方法包括用水泥、環氧植脂的灌漿或噴水泥漿，但必須在有經驗和註冊的結構工程師監督下進行。

Tolerances

Tolerance limits are set up in the specification. The contractor will make sufficient benchmarks and control points to check tolerances.

As a guide, for concrete surfaces, variation from vertical plumb can be 9 mm for each storey. Variation from the horizontal is set to a maximum 6 mm in 6,000 mm. For columns and beans, the tolerance is minus 6 mm and plus 12 mm.

Tolerance is also set up for reinforcement regarding clear distance to formed surfaces, minimum spacing between bars, top bars in slabs or beams at plus or minus 6 mm.

容　限

容限的標準是在施工細則中註明的。承建商須準備足夠校平和控制點來量度容限。

如參考混凝土的表面來說，每層的地台面可以容許 9 毫米距離垂直線；偏差以水平量度計算，可容許 6,000 毫米內有上限 6 毫米的偏差。以柱及樑計算，容限是負 6 毫米及加 12 毫米。

鋼筋工作中亦有註明容限的要求：包括表面與鋼筋的距離，最表面的鋼筋距離等，而在地台與樑間的容限都不能超過 6 毫米。

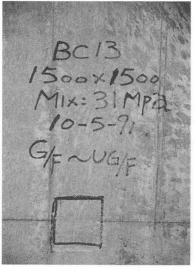

Hammer test on completed concrete

Crushing test

Miscellaneous items

Fair-faced concrete can be produced as an even finish with a sheet material, e.g., plywood or sheet steel, as formwork. Arrangement of formwork panels is specified to form a pattern and texture as desired.

Additions to concrete can be waterproof compound or hardener. The work should be done by a specialist contractor or the materials supplied by a specialist supplier.

The pouring of concrete against a wall of adjoining buildings used as permanent shuttering may result in the failure of portions of such walls and resultant damages. In such cases, independent left-in formwork should be provided to avoid imposed load on adjacent walls, which could not be physically checked for their structural capabilities.

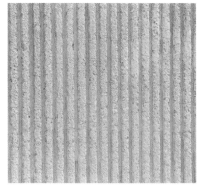

Hammer dressed fair-faced concrete

其他項目

無批盪的清水混凝土可以成為建築物表面的裝飾。模板的物料，包括木板、鋼板等；而橫板的安排組合可以做出各種特別的圖案和質感。

混凝土做好後，再加上防水劑和硬化劑等化合物，就可不用再鋪外牆磚瓦。但是項工程必須由專門的承建商執行，至於防水或硬化劑，亦應由專門的供應商提供。

在隔鄰建築物旁邊灌注混凝土時需特別小心，因為當混凝土貼在鄰近建築物的外牆時，會把部份的壓力附在外牆之上而造成損壞。這樣，便要在灌注混凝土時，在鄰近的外牆邊鋪上並留下模板。

Fair-faced concrete at the Chinese University of Hong Kong

Cutting of finished concrete

Use of fair-faced concrete in façade design

Overall view of building complex with different façade elements

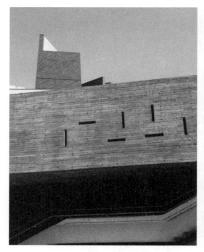
Special pattern and texture achieved by timber planks formwork

Removal of formwork for reinforced concrete wall

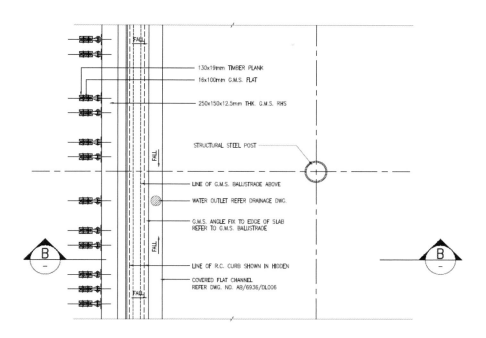

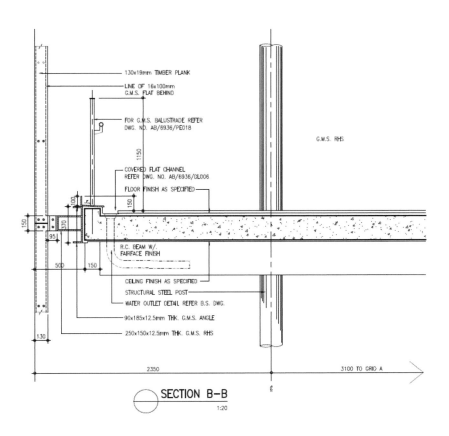

Typical plan and section for balcony

2.4 An Outline of Foundation Systems in Hong Kong
香港常見的地基種類

General foundation design

The foundations of every building function to sustain and transmit safely all the dead and imposed loads to the ground, without impairing the stability of that building or of any other adjacent building, street, slope, or any drainage structure.

The design and choice of foundation will make reference to the following:

1. The soil-bearing capacity of the ground, which is dependent on the type of soil as determined by the site investigation. Data from adjoining sites can be used as reference. The level of bearing strata thus established determines the level of the foundation and if a deep or a shallow foundation is desired.
2. The probable differential settlement of the building and the probable total settlement of the building.
3. The surrounding environment overloading the foundations of adjacent buildings or the ground supporting such foundations needs to be avoided by dewatering or vibration or direct loading distribution. Interference with drains, nullahs, sewers, or other services must be prevented. There must be no unstable conditions created in adjacent slopes or roads.

普遍的地基設計

每幢建築物的地基設計，都是用來傳遞和支撐建築物本身及承受的重量致安全到達地下，亦不會影響鄰近的建築物、街道、斜坡和渠務結構。

設計和選擇合適的地基類別是受下列因素所影響：

(1) 泥土的承重能力，指地面下的土壤種類，通常會在地質勘測報告中列明。鄰近街道的情況亦需用作參考資料。找出承重的地基層才能設計地基的深度。
(2) 承受建築物重量的土地不均勻沉降。
(3) 因為抽水、震動或重量分佈，而引致鄰近建築物的土地荷載超重；必須避免影響地基支撐力的情況，以及影響公共渠務、污水管或其他地下設施。

Foundation types

According to the depth of the bearing strata, foundations can be classified as shallow or deep. Generally, shallow foundations go to approximately 5 m and support a load of 50 to 300 kPa. Shallow foundations include spread footing, strip footing, raft foundation, and grillage foundation. Deep foundations include piling and bored piles. These require a pile cap or caisson cap to transfer the load from the superstructure to the foundation.

Special foundations can make use of the diaphragm wall way of construction to construct the basement floor.

Different foundations may require different methods of testing, which will take up considerable time before the next stage of construction work such as pile cap or superstructure. The construction programme should cater to such necessary timing.

地基種類

地基因承重的地下層深度而有「深」和「淺」兩個類別。一般而言，淺地基的深度可至 5 米，承受 50 至 300 千帕（kPa）的重量壓力。淺地基包括展式基礎、條型基礎、筏基和格床基礎等。深地基的設計有樁基礎、螺旋樁等，其設計原理是藉著樁承台把上蓋的重量移至地基。

特別設計的基礎，包括利用地下連續牆（隔牆）的建築方法來建造地庫。

不同的基礎有不同的測試方法，每種方法都有不同的時間要求。故此，應計算所需的時間並記錄在施工的預算表中。

Site investigation work

Strip footing in construction

Shallow foundations

Spread footing is a reinforced concrete mass supporting the point load from columns and can be classified as rock or soil footing.

Strip footings are also reinforced concrete mass supporting load-bearing walls.

A raft foundation is a continuous slab of spread footings used on poor soil to acquire a big base to spread out the loading of the superstructure.

A grillage foundation makes use of a grid system of H-section steel beams to spread out the load. Concrete is then used to cover the steel beams. This method is seldom used now.

淺基礎設計

獨立基礎是整塊的鋼筋混凝土承擔著由柱所帶著的重量，有時亦稱為石或泥土的基腳。

條型基礎亦是一整塊的鋼筋混凝土承擔著帶重的主力牆。

筏式基礎是完整連續而層展式的，由於有足夠的層面來分擔上蓋的結構和重量，所以用於土質比較差劣的地方。

格床基礎利用格子式的 H 形工字鐵來分擔重量；混凝土用來覆蓋鋼鐵，但今天已很少用此種設計。

Piling test with imposition of test load

Driving of H-piles

Pile caps

Deep foundations: Piling

Steel piling commonly used for foundations can be as deep as 40 m; 12 m lengths of piles are used and driven by diesel hammer or drop hammer. Additional length is by welding. Common piles are 305 × 305 mm serial size and 110 to 223 kg per metre. Problems created are usually associated with noise, vibration, and the encountering of boulders. Initially a test pile is driven at a site investigation bore hole to ascertain the soil conditions. After all piles are driven, a piling test is carried out with imposition of a test load, twice the design load, on a selected pile. This is maintained for 72 hours, and the total settlement recorded must not exceed 15 mm.

Precast concrete piles such as Daido piles can also be used. These are more economical but cannot take horizontal force due to movement generated by wind. Careful control has to be taken during the driving of piles, to avoid vibration damaging the piles themselves. Sites with too many boulders are not suitable for this type of pile.

Cast in situ concrete piles make use of steel casing driven into soil conditions.

Some timber piles of China fir were used as foundations of tenement houses, but these are now obsolete.

深基礎：樁

鋼樁是香港常見的基礎設計，可達 40 米深。12 米長的樁用柴油樁錘或落錘打樁機打落地下。長於 12 米的樁都是燒焊加長的。常用的樁尺寸有 305×305 毫米及每一米長的重量為 110 至 223 千克。打樁時較容易產生噪音、震盪以及遇到硬石塊等情況。設計地基前要先完成探土報告。利用試樁來進一步證明泥土土質的情況，直到完成該部份的打樁工程之後，再測試樁的實際承重量。測試樁須在 72 小時內承托兩倍之設計重量，而所得的偏差不得多於 15 毫米。

預製混凝土樁如「大度樁」是比較經濟的設計，可用於地基工程，但不能承托橫向的風力，尤其在工程進行時要特別小心，避免因震盪而影響樁支本身。這些預鑄樁並不適用於大石群的地盤。

現場灌注樁利用鋼模鑽探泥土後澆灌混凝土。
杉木樁亦曾用於一部份戰前樓宇的基礎，但現在已成歷史。

Deep foundations: Hand-dug caissons

Hand-dug caissons make use of compression rings of Grade 1A concrete, 75 mm thick and 1 m deep to stand the soil load. Traditionally, the digging of caissons was done by a husband-and-wife team, the husband excavating and concreting the caisson rings and the wife raising and emptying the spoil. Construction is about one day per 1 m ring. Dewatering has to be carried out if water is encountered during excavation, and the effect on adjacent buildings is to be monitored. The depth of the caisson can go to 40 m. The diameter depends on the design and may range from 1.5 m to 7 m where excavating machines can be used for the digging. When excavation is completed, pre-fixed reinforcements are put down into the caisson and concreted.

After concrete has acquired 28 days' strength, core drilling is carried out for the caisson. Full depth is not less than 600 mm into the ground upon which the caisson is founded. The core is then taken for crushing tests by an independent testing authority.

Records of soil layers excavated and dimensions and materials for the caissons are kept and submitted to the Building Authority.

This kind of foundation has been banned in 1995 because of high risks for workers. However, this method in the old days provided very good visual observation of soil and anchor rock quality.

Concreting of caisson

Rock excavation by mining

Steel formwork caisson rings

人工沉箱

以手挖掘的沉箱工程，是利用 1A 級混凝土製造的受壓環（75 毫米厚及 1 米深）去承擔泥土的重量。傳統的挖掘工作多由夫妻夥伴一組，男工負責深入地底進行挖掘和澆灌混凝土環，而女工則負責清理挖掘出來的泥土。如工程進展順利的話，平均每天可以完成一米深的環。同時，遇有地下水滲入，必須以適當的抽水方法清理，更要注意鄰近建築物的結

Hand-dug caissons

Digging of caissons

Batching plant of bentonite

Drilling

Removal of soil

Concreted pile

構會否因抽水工程而受影響。沉箱的設計可深至 40 米。直徑則與本身的工程設計有關，一般而言，由 1.5 至 7 米。挖掘工作有時需要機械的協助。挖掘完畢後，把準備好的鋼鐵籠放進洞內，再澆灌混凝土，便告完成。

當混凝土達到 28 天的固定強度，便可施行核心鑽探測試來確認結構的標準，由沉箱地面向下不少於 600 毫米的深度作測試，核心條需由合資格的獨立測試中心進行壓碎測試。

土壤的挖掘報告和物料的詳細資料，必須保存妥善，以便將來呈交建築事務監督作紀錄。

現在因安全問題在 1995 年已禁止使用人工沉箱，但它的好處是能以肉眼視察每一段挖掘泥土的質素及基礎石的好壞。

Deep foundations: Bored piles

Large-diameter (over 2 m) bored piles are machine-dug caissons where steel casing is drilled into the soil, which is then removed. Reinforcement is laid and the caisson concreted. Testing is carried out in the same way as for caissons. These bored piles are used for areas where vibration and dewatering are not desired. However, the site should be large enough for such an operation.

Small-diameter bored piles are similar but use H-steel as reinforcement. These are often used to replace steel piling when the latter encounters boulders.

深基礎：鑽孔樁

直徑較大（超過 2 米）的鑽孔樁，需用機械式的沉箱工程。鋼鐵樁形成樁的內圍，因此可以同時將圍內的泥土清理。鋼筋籠可以安在挖妥的地方，澆灌混凝土再進行定時測試。不過，鑽孔樁常用於較大的地盤，因為這些地方不用抽水以及能避免震盪等情況。

直徑較小的鑽孔樁建築工程跟前述的沒有分別，但鋼筋鐵會以 H 形的鋼鐵板取代。有時鋼樁遇著堅固的大石，亦會採用此類設計。

2.5 Brickwork and Blockwork
磚塊

Materials

Brick or building block is made up of hard, well-burnt clay, natural or cast stone, concrete, and other incombustible material of similar hardness and durability. Resistance to crushing is listed as below.

Cast stone and concrete blocks are cured at normal temperatures and not used for four weeks.

The nominal size of clay bricks is 225 × 105 × 70 mm. Concrete block is constructed of mixes as one part Portland cement, two parts clean washed sand, three parts granite fines, plus five parts 10 mm granite aggregate. Standard block size is 390 mm long by 190 mm deep with varying thickness to suit wall construction. However, actual size varies a lot.

物料

建築工程中的磚塊都是由堅硬的燒面泥、天然或人造石、混凝土與不助燃的物料組成，而且必須有同樣的硬度和耐用性，能抗壓碎等，包括如下：

Brick or building block
建築磚塊

Location 位置	Solid or Hollow 通心/實心	Resistance to crushing in MPa of gross horizontal area 每平方米的受壓度 (MPa)
External or internal (load bearing) 內部或外部的受力重量	Solid 實心	10 10
External or internal (load bearing) 內部或外部的受力重量	Hollow 實心	5 5
External (panel) (non-load bearing) 外置 (板模) (非受力結構板)	Solid or hollow 通心/實心	3.5 3.5
External (partition) (non-load bearing) 內置 (板模) (非受力結構板)	Solid or hollow 通心/實心	1.5 1.5

一般的人造石和混凝土磚會在普通溫度下凝固，需要待四個星期之後才可以使用。

黏土磚的普通體積為 225 × 105 × 70 毫米。而混凝土磚則由 1 份的卜德蘭水泥，2 份潔淨的水洗沙，3 份的麻石碎粉混合在 5 份的 10 毫米麻石粒中，一般的混凝土磚體積為 390 毫米長、190 毫米深，厚度方面隨著建築的需要而改變。

Traditional blockwork: Roman aqueduct at Segovia, Spain

Traditional blockwork: Roman aqueduct at Segovia, Spain

Strength and shrinkage of concrete blocks

For block work not less than 75 mm thick, the average crushing of ten blocks will be no less than 2.8 N per mm^2 and for individual block, its strength should not be less than 80% of the average.

The average drying shrinkage of a sample must not exceed 0.06% (BS 6073), except for aerated concrete blocks.

混凝土磚的強度與收縮性

不小於 75 毫米的混凝土磚，每 10 件的平均受壓度應不少於 2.8 / 平方毫米。而每一磚塊受壓度亦不能少於 80% 平均數。而物件的平均乾涸速率，要以不超於 0.06% 為原則。但輕磚並不在此規限中。

Mortars

For general brickwork and blockwork, cement mortar is used. Cement mortar is in a ratio of cement to sand of 1:3. Cement is ordinary or rapid-hardening Portland cement, usually delivered in bags to the site. Sand consists of naturally occurring clean, hard sand, durable crushed stone, or both combined. Also, sand contains no greater proportion of fine clay, silt, or fine dust (which will pass through a 25 um sieve) than 5% by mass for sand and 10% by mass for crushed stone. For fair-faced works it is free from salt causing efflorescence.

Cement-lime mortar consists of cement, lime putty, and sand 1:1:6 for external walls and 1:2:9 for internal walls as well as for vermiculite blocks.

Traditional use of external brickwork in Hong Kong

Concrete blocks

Laying of blockwork

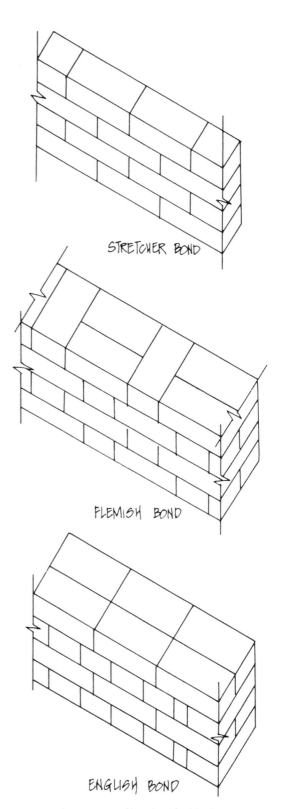

Different types of bonding for blockwork

砂 漿

一般常見的紅磚牆和混凝土磚牆，用的也是水泥砂漿。水泥砂漿的比例為（水泥砂石）1：3。水泥分為普通的或特快凝固的卜德蘭水泥，通常包裝好再運到地盤使用。沙的成份包括天然含量的潔淨的硬沙及壓碎的石子，或者是兩者的合成。同時，沙粒中所含的微粒土和粒子（可通過 25 微米的箕子為準）不能超過 5% 的總質量；而壓碎的石子亦不能超過 10% 的總質量。用於清水混凝土（不經表面處理）的沙不能含有鹽份，以防止風化。

水泥砂漿含水泥、石灰膏和沙的比例為：

(1) 1：1：6 適用於外牆建築；
(2) 1：2：9 適用於內牆和混凝土磚的組合。

Workmanship

In dry weather, bricks and blocks are just wet enough to prevent premature drying out of the mortar. In wet weather, freshly laid work is protected during interruptions from rain and at the completion of each day's work.

To ensure stability of the structure and to produce a better appearance, the laying of bricks is done in a recognized pattern or bond, such as a stretcher bond, an English bond, or a Flemish bond. These patterns are planned to give the best practical amount of lap to the bricks, which should not be less than a quarter of a brick length.

Bricks or blocks are laid on a full bed of mortar with the joints filled solid to a consistent thickness of 10 mm. Work is carried out with no portion more than 900 mm above another at any time, racking back between levels. In facing or fair-faced work, bricks are completed up to the level in one operation. The brick courses are kept level and plumbed at all wall faces, angles, and features.

For infill brickwork or blockwork, a 300 mm wide metal lath is installed across the joint between the brickwork or blockwork, and concrete before plaster is applied to prevent future cracks along the joint.

Tolerance for blockwork is set at ± 10 mm for setting out dimensions and vertical plumb in any storey height.

施 工

在乾燥的天氣下，磚和混凝土磚都必須保持濕潤，以防止砂漿還未凝結時便乾涸。但在比較潮濕的天氣下，剛剛完成的工作都要小心蓋好，防止驟雨的影響。

為了保持結構上的穩固和外觀的原因，磚頭都是有固定和規律的排列方法，包括順砌法、英式或法式砌法。施工時，磚與磚的重疊距離不能少於四分之一磚塊的長度。磚頭是排放於砂漿的底，接連的空間要塗滿 10 毫米厚的砂漿。施工過程中，每一次鋪砌的高度不能多於 900 毫米。

在鋪砌見光的磚牆時，磚頭要以同一個工序完成鋪砌，同時要保持垂直及平面標準，亦要保持劃一的定角標準。

在施工的內牆，需先在混凝土與磚塊介面擺放 300 毫米的鐵網，然後做砂漿批盪，來防止在介面發生裂痕。磚牆開線的公差距離為 ±10 毫米。

Requirements for mortar (BS 5628: Part 1:1978)
水泥標準要求

		Mortar designation	Type of mortar 水泥類別 (proportion by volume 體積比例)			Mean compressive strength at 28 days 28天的平均壓力	
			Cement: lime: sand	Masonry cement: sand	Cement: sand with plasticizer	Preliminary (laboratory tests)	Site tests
Increasing strength ↑↓	Increasing ability to accommodate movement, e.g., due to settlement, temperature, and moisture changes					N/mm²	N/mm²
		(i)	1:0 to ¼:3	–	–	16.0	11.0
		(ii)	1:½:4 to 4½	1:2½ to 3½	1:3 to 4	6.5	4.5
		(iii)	1:5 to 6	1:4 to 5	1:5 to 6	3.6	2.5
		(iv)	1:2:8 to 9	1:5½ to 6½	1:7 to 8	1.5	1.0

Direction of change in properties is shown by the arrows

Increasing resistance to frost attack during construction →

Improvement in bond and consequent resistance to rain penetration ←

Cavity wall

A cavity wall is constructed in two leaves with a space or cavity between them. The two leaves are constructed of solid bricks or building blocks, each not less than 100 mm thick and the intervening cavity not less than 50 mm and not more than 75 mm wide. Galvanized iron ties not less than 20 × 3 mm in cross-section are used to unite the two leaves but not to transmit moisture. Such ties are built into the horizontal bed joints at distances not exceeding 900 mm horizontally and 450 mm vertically apart. During construction, the cavity must be kept free from mortar droppings.

Though occupying more space, cavity wall construction has good weather protection and better thermal and acoustical insulation than a standard one-brick thick wall. However, rigid site supervision is required to ensure good workmanship.

空心牆

空心牆是在兩堵牆的中間預留空間。牆由實心磚或空心石塊所建，但其厚度不能少於 100 毫米，而其中空間的距離亦不能少於 50 毫米或多於 75 毫米。連接兩道牆一起的鍍鋅鐵板最小的切面為 20 × 3 毫米，可防止濕氣滲入。連接鐵板須於施工時放於橫面的介面上，橫向距離在 900 毫米之內，垂直的距離亦為 450 毫米之內。施工期間，中空的部份要保持清潔，不能溜進英泥或砂漿等。雖然空心牆的設計較浪費空間，但它的優點是較一般磚牆更具隔熱和聲控的性能。不過，空心牆的施工技巧要求較多的質量監察。

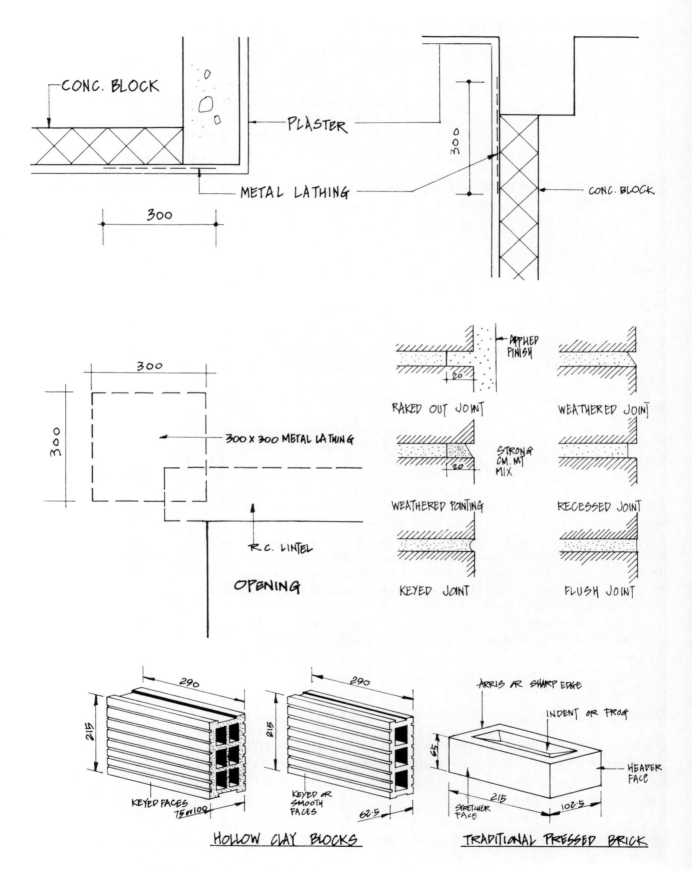

Blockwork: joint between block and structure, external joint types, blocks, and bricks

Limitations

Every external wall of a building should be constructed of:

1. masonry not less than 225 mm thick;
2. plain concrete or reinforced concrete not less than 100 mm thick;
3. any of the materials mentioned in paragraph (1) or (2) in combination with a framework of steel or reinforced concrete; or
4. other suitable materials of permanent, non-combustible and impervious construction.

Generally, for a boundary wall or fence not exceeding 1.8 m high, brickwork or blockwork must not be less than 100 mm thick. If the wall or fence exceeds 1.8 m high but less than 3 m high, the wall thickness must be at least 225 mm. Buttresses or piers not less than 225 mm² in plan and not more than 2 m apart, centre to centre, are constructed and placed at all angles and ends of such walls.

For fire rating, solid bricks of clay, concrete, or sand lime without plaster of 100 mm give one hour FRP; those of 225 mm give two hours.

限 制

建築物周邊板牆應根據以下指引承建：

(1) 用石塊作牆身厚度不少於 225 毫米。
(2) 混凝土或鋼筋混凝土作牆身厚度不少於 100 毫米。
(3) 鋼筋或鋼筋混凝土框架配合以上 (1) 或 (2) 所述之物料。
(4) 其他永久性、不燃性及不透性之適當材料。

根據一般界定在不高於 1.8 米的界線牆和圍籬的設計中，磚或空間石塊須超過 100 毫米的厚度。介乎於 1.8 至 3 米的高度時，牆身的厚度應為 225 毫米，設計的撐牆或支墩要有 225×225 毫米的切面，兩支墩之間的距離由中心至中心須有至少 2 米。

在防火標準限制下，100 毫米的黏土磚、混凝土磚或砂磚可以提供一小時的防火設計；而 225 毫米的，可達至兩小時的防火時間。另外，75 毫米的混凝土已有一小時防火；而 100 毫米的則可高達兩小時。

Random rubble

Square rubble

Ashlar stonework

Square coarsed rubble

2.6
Masonry and Granite/Marble Works
石工、雲石和麻石

Traditional masonry

Traditionally, local granite from Hong Kong or China was used as the stone for masonry. The application is for facing at external walls, fence walls, retaining walls, and pavings.

Rubble walling consists of stones finished in a rough or uneven surface, presenting a natural appearance. Patterns of stones include random rubble, square rubble, and square coursed rubble. The thickness of stone is 75 mm minimum, and mortar joints are 5 to 15 mm wide.

Ashlar walling consists of stone finely worked to a plane surface and generally laid in a square or grid pattern. A mortar bed is composed of cement and fine crushed stone 1:3 with 5 mm joints.

For stone walling constructed as facing to concrete or brickwork, at least 5 wall ties per square metre are built in for a depth of 100 mm and to project 75 mm into the stone walling.

Stone pavings are built with a thickness between 50 and 150 mm. For large slabs, a bedding of dry sand is required. For smaller units, a bedding of 1:3 cement and sand mortar is used. Joints are grouted with a similar mortar prepared with crushed stone to match the paving.

Traditional masonry has a unique texture and character, but due to the intensive labour and the high cost required to prepare the stones, it is not much favoured nowadays by Hong Kong architects. However, modern granite cladding prepared from a high-tech factory is becoming more widely used in architectonics revitalizing the masonry tradition.

傳統石工

傳統上，本地的麻石都是產自中國和香港，常用於鋪砌外牆、周邊圍牆、擋土牆和地面。一塊石牆是由石頭組成，經表面處理後變成不規則、表面粗糙，有如天然的麻石。石頭鋪砌圖案的包括：

(1) 不規則；
(2) 方塊鋪砌；
(3) 方塊碎狀。

石頭的厚度不能少於 75 毫米，並要有 5 至 15 毫米的水泥介面。

方石施工用細緻而表面平滑的石塊合成，通常採用正方形定格的圖案。水泥底床由水泥和微細的石合成，比例為 1：3，連縫間為 5 毫米的距離。

The Luk Kwok Hotel façade and colonnade

The Bank of China landscape

The Standard Chartered Bank bridge

Modern granite (the Hong Kong Club façade)

The Hong Kong Convention and Exhibition Centre entrance

The Bond Centre cascade

　　在混凝土或磚牆外鋪砌石塊作粉飾也有一定的程序，每平方米至少要有五個連接力牆的支撐結構，同時須留有 100 毫米深度於外牆和 75 毫米的部份藏於石塊之內。

　　以石片鋪砌地面的厚度分為 50 至 150 毫米。在鋪砌較大的石片時，要用乾沙作底層基礎；而基礎底由水泥與砂漿以 1：3 合成時，可用於較薄的石面。所有連接介面都要用相若質料的灰漿和碎石填塞。

　　傳統的砌石工作可塑造獨特的質感和格調，但往往需要比較多的工人，很自然地提高了建築的成本，所以香港設計師一般都較少用砌石裝飾。但是，隨著現代科技的發展，由工廠加工和製造的新的外牆覆蓋層，亦能表現麻石和石片的鋪砌效果。

Granite assembly in Vancouver, B.C.

Stone panel integrated with aluminium honeycomb panel as option for lightweight use at external wall

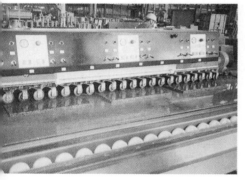
Polishing

Cutting by disc saw

Modern marbles, granites, and stones

The 1980s witnessed the retrieval of stone technology transformed by high-tech machinery to suit the style of post-modernism induced by American and European architects. On the external wall, the general design technique is to contrast with large glazing or a curtain wall to create a strong textural difference. Granite slabs can also be combined with glass as part of the curtain wall system or cladding system. Advanced technology can also afford the refined production of granite ornamental work for buildings such as mouldings, architraves, sculptures, and other architectural features where magnificence, luxury, and splendour are required.

現代的雲石、麻石技術

八十年代見證了高科技革新，並運用於石塊技術上，以適應後現代主義風潮下歐美建築師新概念的建築。外牆上往往以透明的玻璃面來塑造強烈的對比，豐富表面的質感變化。麻石片塊與玻璃一併運用，以麻石的重量質感混著玻璃幕牆的技術，成為外牆的一個部份。先進的科技發展，製造出許多誇張和美觀裝飾用的石線與版面圖案，增加了外牆設計的變化。

Production

The quarrying of stones is done in Italy, the US, China and many countries, providing a variety of colour and texture for every purpose. The quarried stones are kept in a covered or an open stockyard, depending on the type of stone.

The cutting of the stones is by gang saw or disc saw into the required slab thickness, after which the stones are sanded by machinery to a plain surface: honed finish. For granite works when a rough surface is required, a flame is set to burn out some of the quartz on the granite surface: flamed finish. A smooth, even surface can be obtained by polishing the stones after cutting: polished finish.

There are also special machinery to cut, grind, or polish different stones into various shapes, a common one being the polished circular column cladding formed by assembly of the stones on the factory before polishing.

Mock-ups can also be built for examining the final appearance of the stones. Experience shows that large panels set up outdoors give a good sample before the right kind of stone can be chosen and specified. As a natural material, the stone is sorted out in batches of the same colour for different areas of application.

生 產

麻石或雲石的開採與加工，多數在義大利、美國和中國，其他地區亦提供很多不同類別、顏色及質感的石。從石礦開採出來的石頭會視乎種類，貯藏於遮蔽的地方或空曠的礦場中。

石片由鋸床或碟形鋸扇切開成需要的厚度，然後運到磨砂機打磨，成為「平啞面麻石」。麻石粗糙的表面須放入火爐燒去面層的石英，成為「燒面麻石」，再由磨砂機磨成平滑光亮的表面，即「光面麻石」。

此外，較特別的形狀，如光滑圓柱鋪面板，則要以特別的機器來製造；板面要在工場內先裝嵌成形，才開始進行打磨或切開。

在鋪砌石片過程中，往往需要預先裝嵌樣板作為測試。經驗證明用較大片的石板作樣板較為準確。由於天然石的成份會有參差的情況出現，故在鋪石片之前，應先於同一批石材內選擇顏色相同的石片，可避免外觀的參差。

Grinding

Stone cladding

Cladding means an external facing or architectural decoration in addition to the external walls of a building. Sufficient permanently flexible joints are provided horizontally and vertically to allow for differential movement in the cladding and in the attached structure. Metal dowels and fixings securing the cladding are permanent and corrosion protected. Generally, sand or cement bedding alone is considered unacceptable as a suitable fixing method. The plan for the external cladding should be shown in the general building plan. For cladding above 6 m from ground level or the adjoining floor, the Building Authority requires details of the thickness, strength, durability, fixings, and sequence of support to be submitted for approval in the structural plans.

The actual thickness of granite slabs for cladding depends on the type of stone and the size of panels, generally ranging from 30 mm to 40 mm. A 1000 × 1000 × 30 mm is a good size to handle. The granite slabs are fixed by stainless steel hanger, bolt mounted, say, 40 mm, into concrete. A space of 65 mm to 80 mm is left between the external wall and the cladding, to allow for drainage and tolerance in the concrete structure. A waterproofing coating on top of the concrete surface is preferable. Slab joints can be open or sealed with silicone at 5 mm to 10 mm wide.

Cutting by gang saw

Nowadays, backing materials, like honeycomb panels, are used to enhance the performance of stone. The composition of a panel will be 4 mm to 6 mm natural stone veneer, aluminium honeycomb of 13 mm to 18 mm. The total thickness is around 20 mm to 25 mm as calculated to meet the strength requirement. This product is more lightweight than natural stone is and stronger in strength and anti-impact. Also, easier and quicker installation is one of the most significant advantages. Different treatment of stone can also be applied to the stone veneer to introduce a different texture from that of normal natural stone.

Flaming

As stone is a natural material, the physical properties and chemical properties can vary significantly from different batches and types of stone. There are specified on-site tests on anchors and panels to be carried out to verify the performance and workmanship of the anchors installed.

石的裝嵌

外牆裝嵌是在建築物外牆加上一些設計去裝飾表面。足夠的伸縮性接駁結構，需要在橫直兩個方向上安裝妥當，使石表面受因溫差影響後可自由擴張和收縮。把表面的石板安全地扣緊於牆身外的鋼鐵含釘和碼子，須具有防鏽和永久性的作用。一般而言，單是沙石泥水底是不能被視為合規格的石片安裝方法。覆蓋層工程之圖紙必須展示在一般建築圖則上。地面或比鄰層高 6 米以上的覆蓋層，建築事務監督在批閱圖則時，

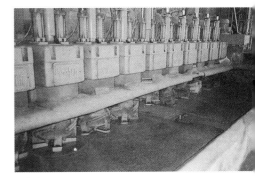
Polishing

也往往要求建築商提交有關的安裝方法，經批准後才可以安裝在結構圖則上。

麻石片的實際厚度要視乎石的類別、板片的形狀和大小而決定，一般介乎於 30 至 40 毫米。1,000×1,000×30 毫米的板較為容易搬運。麻石片由不銹鋼吊片牢固在至少 40 毫米深的混凝土中。片與片之間須預留 65 至 80 毫米的空間，作為疏水和混凝土牆的偏差。此外，在混凝土外牆之表面需塗上一層防水膜。石片之間隙可以採取開放式或密封式的方法，後者可用 5 至 10 毫米矽膠接駁。

現在普遍使用在石材基底的物料，還有蜂窩板，其作用在於能提高石材材料上多方面的性能。複合石材的組成是 4-6 毫米的天然石料表面、13-18 毫米的鋁製蜂窩底，總厚度大約為 20-25 毫米，以符合強度要求。此產品比天然石材更輕，而且強度增加並抗衝擊。此外，簡單易裝是最明顯的優點之一。不同的石材處理方法都可以用於石料表面，以得到不同的天然石材紋理。

因為石材是天然材料，物理和化學變化會出現於不同批次和類型的石材，所以必須通過現場測試嵌固件和板材來核實安裝質素。

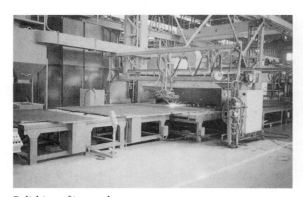

Polishing of internal curve

Assembly for circular cladding

Dry-mounted sample

Polishing

Fixing details

Polishing

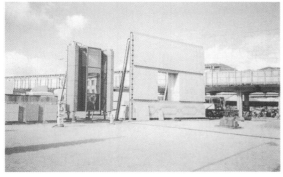
Large-scale sample

Complete circular cladding

Granite tile

Dry-mounted granite cladding

Sekigahara Stone Factory, Japan

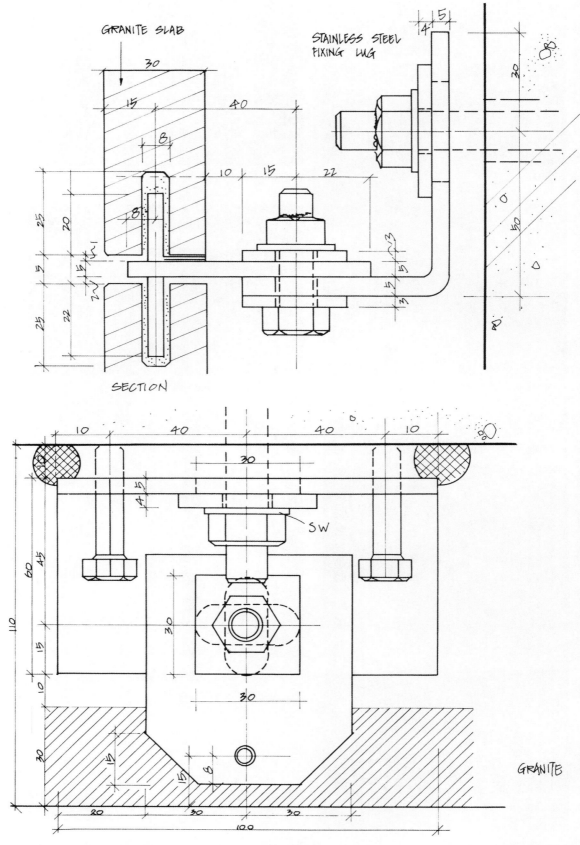

Granite slab : open joint, dry-mount fixing details

Granite slab: open joint, dry-mounted fixing details

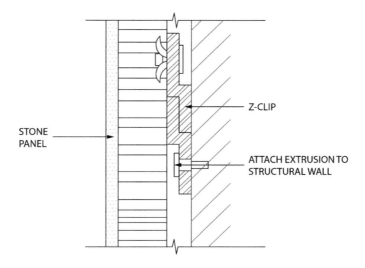

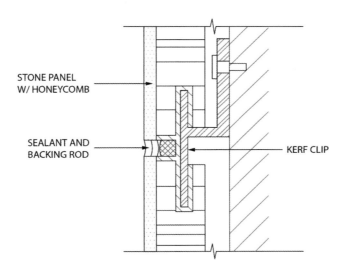

Stone with honeycomb backing: sealant joint, dry-mounted fixing details

2.7
Roofing, Waterproofing, and Expansion Joints
屋頂的鋪工、防水層、伸縮縫

Laying of asphalt

Roofing in general

Every roof must be weatherproof and built with gutters and rainwater pipes to prevent the direct discharge of water on footpaths or roadways. Construction can be of tile, glass, metal, asphalt, or layers of impervious felt membrane.

Accessible roofs are required to be protected by parapet or railings not less than 1,100 mm in height from the finished roof level and constructed to prevent climbing or passage of objects more than 100 mm in their smallest dimension. The lowest 150 mm of roof parapet must be built solid.

一般屋頂處理

每個屋頂都必須作適當的防水設計，還要裝設屋頂邊渠、邊溝並連接雨水渠，以阻止雨水直接流落行人路和街道。施工方法可以用屋頂的磚瓦、金屬、瀝青層等非滲透性物質。

在可到達的屋頂上要裝置高於（以完成的屋頂計）1,100 毫米的保護欄杆或低牆；而且保護欄杆間的距離不能超過 100 毫米，以防物件跌落行人路，造成危險。根據建築物條例，在屋頂面計起需要 150 毫米的實心牆。

Asphalting (roofing and tanking)

Derived mainly from crude oil residue, asphalt is a good material and the most popular in Hong Kong for damp proofing, tanking, and roofing. The work must be executed by a specialist contractor, who may submit a written guarantee that the material defect-free and watertight for a period up to ten years.

Mastic asphalt is composed of suitably graded mineral matter and asphaltic cement in such proportions as to form a coherent, voidless, impermeable mass, solid or semi-solid under normal temperature but when heated to a certain temperature sufficiently fluid to be spread by means of a hand float.

The asphalt blocks coming into the site as hexagonal blocks of approximately 300 × 300 × 100 mm are gradually heated in a cauldron or mixer to a temperature not exceeding 230°C. The heated asphalt is removed by buckets coated with a fine inert dust or cement.

Cement sand (1:3 mix) screed with minimum 25 mm thickness and laid to fall forms the base for the asphalt. Laying is done in bays not exceeding 2 m wide. Successive coats are laid, and joints are staggered at least 150 mm. An angle fillet is required where abutting a vertical surface. Skirtings, covering to kerbs and the like exceeding 300 mm high, are reinforced with metal lathing securely plugged and stapled.

For horizontal roofs, two coats of equal thickness up to 20 mm are applied. For tanking, three coats with a total thickness of 30 mm are necessary.

Until the asphalt is cooled, no traffic is allowed on the surface. This is then rubbed with a clean wood float using fine clean sand. The finished asphalt cannot be used as a building platform or for storing, mixing, or preparing other building materials.

For pipes through roofs, bitumen primer is applied before applying asphalt. An angle fillet and a minimum height of 150 mm are required.

Additional protection against water penetration can be afforded by adding bitumen felt as a vapour barrier before application of the asphalt.

Finishing to the asphalt can be chippings of grey granite or white stone (passing through a 5 mm sieve but retained on a 2.36 mm mesh BS sieve). Reflective paint of bituminous-based bright aluminium paint as a final finish can be used to reflect some of the heat on the roof. However, another way to finish the asphalt is to apply cement sand rendering protection. Then five-legged concrete tile or other tiles can be installed on top to make the finished roof a more useful area.

Thermal insulation can be provided by use of a lightweight concrete screed or by the application of an expanded polystyrene foam board on the asphalt before the cement sand rendering is applied. The five-legged concrete tile or hollow block can also afford some insulation value.

瀝青面（屋頂鋪面及地下層防水面方法）

瀝青是一種由原油油渣提煉出來的物質，也是最好和最普遍採用的防水鋪面，常用於地下層或屋頂上。鋪砌工程須由專業的特別項目經理負責。在監管工程規範中，會要求提交一份十年有效期的保養證明書。

瀝青膠漿含有特別的礦物，混合適當份量的膠脂水泥，可成為一種內在平均的、無空間的和不滲透的物質；在正常溫度下，會呈固體或半固體狀態，但熱度達到某程度時，則會由固體或半固體狀態變為液體狀態，可流鋪於平面上，凝固後成為一層防水的膜。

鋪瀝青前，先要以水泥砂漿（1：3 的比例）做好基本底層，厚度最少要有 25 毫米。鋪盪工作要以不超過 2 米為闊度，重疊的層片需要最少 150 毫米。當橫向的平面接連垂直的平表面，膠漿要呈填角。垂直面如壁腳線、邊石的覆蓋等，如高夾 30 毫米，則需用金屬網加以堅隱。

做平面的屋頂，要有兩層同樣厚度的防水膜，合共不能少於 20 毫米。垂直的平面不能少於三層同樣厚度的膠膜，總厚度最少要有 30 毫米。

瀝青面未凝固前要保護表面，不能踐踏。凝固風乾後，再以清潔的木條和細沙磨擦平滑。瀝青面完成後，不能當作建築平台來貯存雜物或做任何建築物料的預備工作。

如屋頂上裝設有喉管，先塗上瀝青質的表面漆，再在表面塗上真正的瀝青液，而且須有填角和瀝青高度不能少於 150 毫米。

先鋪瀝青紙，才塗上瀝青液膜，可加強防水的功能。

在完成的瀝青表面上，可以鋪砌灰麻石碎或白色石頭（介乎 2.36 至 5 毫米）。反光的瀝青並含有鋁質金屬油漆，可作為最後的表面處理，而且亦有反射熱能的作用。另外的處理方法是，於表面上先塗上水泥砂漿，再鋪砌「五腳」的混凝土磚塊，屋頂便可承重。

至於隔阻熱能的方法，可用輕盈的混凝土漿，或在水泥砂漿上加伸縮性的碳纖維板，但「五腳」混凝土磚塊和空心塊板亦有隔熱的效果。

Composition by analysis of mastic asphalt: BS 988 (Roofing) and BS 1097 (Tanking and Damp-proof Courses)
地瀝青膠漿的構成（地層防水和防水鋪面）

Property	Percentage by Weight of Mastic Asphalt			
	BS 988		BS 1097	
	min.	max.	min	max.
Soluble bitumen	11	13.5	12	15
Passing 75 um mesh BS sieve	35	45	38	50
Passing 212 um mesh BS sieve and retained on 75 um mesh BS sieve	8	22	8	26
Passing 600 um mesh BS sieve and retained on 212 um mesh BS sieve	8	22	8	26
Passing 3.35 um mesh BS sieve and retained on 600 um mesh BS sieve	12	23	4	17
Passing 3.35 um mesh BS sieve	0	3	0	2

Bitumen felt built-up roofing

Bitumen felt consists of one layer or more of fine granule-surfaced underlay, a top layer of mineral-surfaced underlay, and a top layer of mineral-surfaced felt or fine- granule-surfaced felt.

On a concrete roof, a cement sand screed makes up the base. Then a coat of bitumen primer is brushed on and allowed to dry. The first bitumen felt layer is partially bonded to the base at the perimeter and in spots or strips, with hot bitumen-based bonding compound at the rate of 0.5 kg per m^2. Subsequent layers are then fully bonded with an even coating of hot bonding compound at the rate of 1.5 kg per m^2. The final surface is dressed with bitumen compound applied at the rate of 3 kg per m^2 dressed immediately with stone chippings lightly rolled in at the rate of 15 kg per m^2.

Asphalt blocks

瀝青紙的屋頂保護

瀝青紙共有三層，先做好一至二層細沙面的底紙，最後一層以礦物質面或細沙質面紙完成。

在混凝土的屋頂上，先由水泥砂漿完成一層基本底層，隨後鋪上第一層的瀝青紙，待乾涸後，再以 0.5 kg/m^2 的速度塗上另一層的熱瀝青漿。第一層的瀝青紙首先接緊基礎層的水泥砂漿，第二層的瀝青以 1.5 kg/m^2 的密度緊扣第一層的瀝青面。最後一層的瀝青再以 3 kg/m^2 的密度灌注，同時以 15 kg/m^2 的密度混進碎石於表面。

Wire mesh for vertical reinforcement

Bituminous emulsion roofing

Bituminous emulsion roofing is applicable to concrete flat roofs, treatment of cracks in existing felt covered roofs, treatment of joints in steel corrugated sheet roofing, etc.

The bituminous emulsion is to be used in accordance with the manufacturer's recommendations. For concrete surfaces, a wash coat of bitumen emulsion and water 1:1 volume mix is first applied. Then a priming coat of bitumen emulsion and water 5:1 volume mix is applied. After one coat of emulsion at a rate of 0.7 litre per m^2, one layer of glass fibre membrane is installed, and three coats of bitumen emulsion at 0.5 litre per m^2 per coat follow.

For cracks in existing felt-covered roofs, first the wash coat and primer are applied. Then bitumen emulsion and sand 1:3 by volume is used to fill the cracks. The entire roof is then coated with the bitumen emulsion.

Laying on roof slab

瀝青液屋頂

這種施工方法適合於一般平面屋頂，主要是處理先前防水層的裂縫或鋼鐵天面接駁的地方等。

瀝青液的使用方法，需依據供應商的指示。處理混凝土面時，洗水瀝青漿和水（1：1 的比例）要先塗抹，接著是 5：1 的瀝青漿與水的混合。再造一層每平方米 0.7 公升的瀝青液，再加一層玻璃纖維及以每平方米 0.5 公升的密度灌注三層的瀝青液漿。

在處理屋頂瀝青紙的裂縫時，先以洗水層和基本層鋪上，再以 1：3 的比例混合瀝青液和砂鋪滿在裂縫上，最後以瀝青灌注填平整個屋頂。

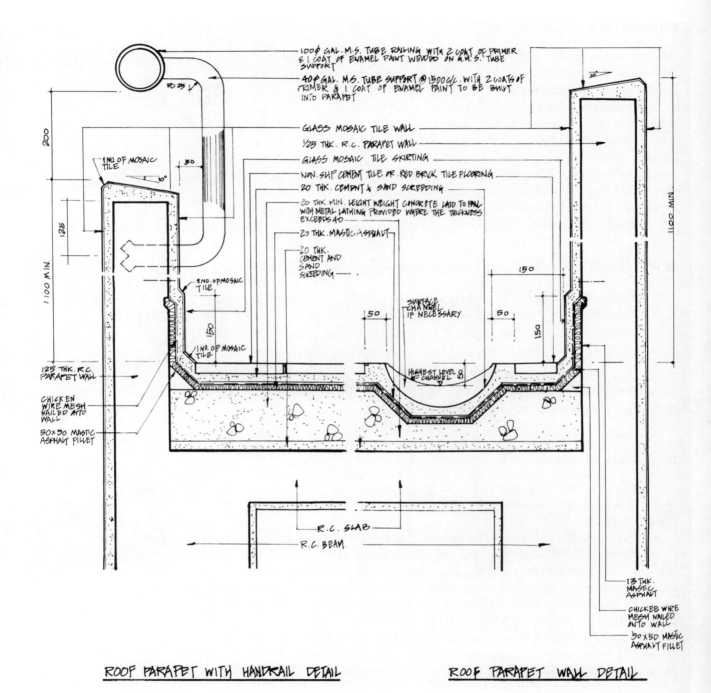

Roof parapet details with handrail and roof parapet wall detail

Clay tile roofing

Tiles are made from clay and concrete in a wide range of designs and colours. Plain tiling is laid so that at least two thicknesses of tile covers every part of the roof and bonded so that no 'vertical' joint is immediately over a 'vertical' joint in the course below.

Another tiling method is single-lap tiling with overlapping side joints. A common form is the pantile, which has opposite corners mitred to overcome the problem of four-tile thicknesses at the corners.

Due to strong typhoons in Hong Kong, clay tile roofing is employed more often as a decoration over a concrete roof already installed with waterproofing.

黏土瓦屋頂

由黏土和混凝土做成的屋頂瓦片有許多不同的顏色和形狀。鋪砌瓦片的時候，要確定每塊瓦片都重疊在先前鋪砌的瓦片上，這樣才使到至少兩砘瓦片重疊，及垂直介面不會在同一位置。

另一種鋪砌屋頂瓦片的方法，是單疊式的鋪砌法，以重疊瓦片邊旁成為介面。另有特別設計的「瓦片」，主要用於轉角的位置，避免遇有四層瓦片同時重疊的情況。

香港的地理位置，經常受到颱風吹襲，因此以黏土瓦鋪砌屋頂只用作裝飾。一般的設計是在混凝土層上加上防水保護，再以黏土瓦鋪好表面。

Clay tile roofing

Concrete roofing

Copper roofing at No. 9 Queen's Road Central

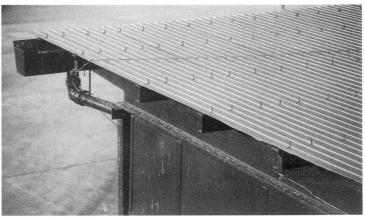
Corrugated steel sheet for a temporary refuse collection chamber

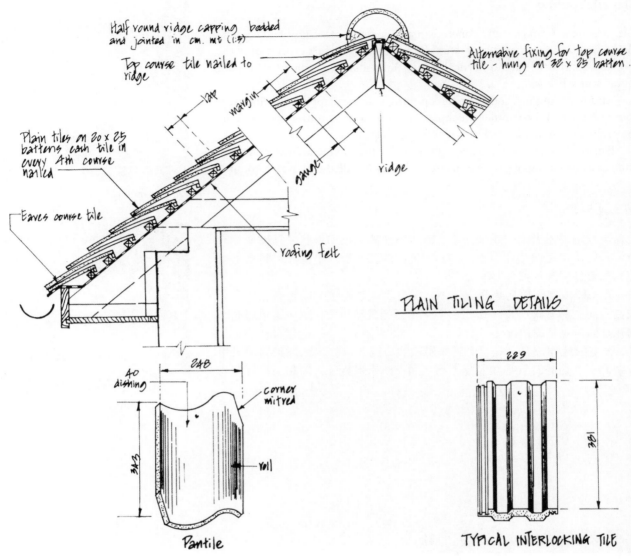

Plain tiling details

Metal roofing

Metal for roofing includes corrugated steel sheet, corrugated aluminium, milled sheet, and rolled copper sheet, strip, and foil.

Steel corrugated sheets are hot-dipped galvanized, at least 0.6 mm thick with corrugations 76 mm wide and 19 mm deep. Fixing is by self-tapping galvanized steel screws and bolts. Aluminium corrugated sheets are similar to steel sheets.

Lead and copper are traditional roofing materials used for thousands of years in Europe. For lead, an underlay of felt or building paper is required to isolate lead from any free lime present in the screed. For copper, a bitumen coat of underlay will prevent any moisture present forming dilute acids which may react with the copper. However, neither lead nor copper is a traditional roofing system in Hong Kong. They are therefore rarely used.

金屬面屋頂表面

以金屬作為鋪砌屋頂的材料，包括竹節的鋼鐵片、鋁片、鉛片和軋制銅片，以及銅條和箔。

波紋金屬屋面通常要經熱能電鍍的處理，鋼鐵的厚度不能少於 0.6 毫米，而波紋的部份要有 76 毫米寬和 19 毫米深。在接駁兩塊瓦片時，要確定兩塊瓦片間有重疊的位置，在上面加鉛水鐵釘作穩固。安裝鋁金屬竹節片的方法，大同小異。

歐洲以鉛和銅作屋頂物料，已是好幾個世紀以前的事。以鉛來說，首先在沙石底層上，安妥一層油紙作為分隔鉛片和沙石。至於銅片屋頂的處理方法，是先以瀝青紙作為底層加於沙石面上，作用是以瀝青紙為防濕層，避免酸性成份直接滲入銅片之內產生化學反應。但此兩種屋頂的處理方法，並非香港傳統工序，所以較為罕見。

Waterproofing (in addition to roofing)

The type of waterproofing depends on the nature of the building or structure, whether the water is under pressure or not, the access available, and the construction materials. General requirements can be thus categorized:

1. External waterproofing above ground using either waterproof concrete or an external waterproof rendering and screeds.
2. External waterproofing to the existing system under water pressure as water retaining structures, e.g., groundwater reservoirs or water tanks.
3. Internal waterproof renderings for below ground structures, such as basements, lift shafts, and swimming pools.
4. External waterproofing to new construction below ground, e.g., foundation and basement.
5. External vertical surfaces requiring a water-repellent surface but not strictly waterproof.

The waterproofing materials available are also thus categorized:

1. Integral waterproofers as additives to cementitious mixes, e.g., concrete, mortars. These operate as pore fillers combined with water repellent chemicals.
2. Pre-packed integral waterproofers requiring only the addition of water at site, which makes site mixing simple. These contain active chemicals reacting with water and free calcium in the substrate to form insoluble crystals which block the capillaries, preventing the ingress of water.
3. Polymer-modified cementitious systems as emulsions when used in cement sand mixes; the emulsion breaks down to form a fibrous lattice within the mortar structure, affording improved adhesion, flexural strength, and reduced permeability.
4. Emulsions based on bitumen or bitumen or rubber designed to provide either waterproof external coating or waterproof membranes. These should be covered with a screed for protection.
5. External water repellents based on silicone resins or stearate complexes designed to repel rain from permeating vertical surfaces. However, these allow the release of dampness from within the substrate.

防水處理（屋頂以外）

防水的類別及方法，主要取決於建築物的用途和結構，還有水壓、方位和建築用料等因素。

一般要求可分為下列幾個要點：

- 地面以上的外面防水設計，採用防水混凝土或表面塗上防水抹灰及底層灰漿。
- 表面處理的防水工程，用於現有受水壓影響的處境，如需要儲水的結構或地面的水庫等。
- 內部的防水底層抹灰，用於保護地面以下的結構，如地庫、電梯槽、泳池等。
- 表面處理地面以下的新結構，如地基工程或地庫設計。
- 需要避免多水浸透的垂直表面處理，但並非要求全面性的防水設計。

至於防水用料，亦可概括為下列幾項：

1. 完整性的防水，附加於水泥混合物中的溶劑，如混凝土、泥漿等。它們含有微細的空隙作濾網，且與抗水化合物產生化學作用。
2. 獨立包裝的完整防水劑，一般在地盤裏加水混合而成。它們含有活性化學成份，一旦與水和當中的鈣份子結合後，便會產生化學作用，形成非水溶性的晶體來阻隔水從微絲管的導口浸入。
3. 碳合成物與水泥砂漿混合時所簡化了的水泥成份，使它們再結合，重組成纖維性的網體，從而增強它的黏性、彈性來減低滲透作用。
4. 以瀝青、橡膠為主的乳劑，可用作表面的防水處理，但須在底層加上抹砂工夫才算完成。
5. 表面阻水的物料，以矽脂或硬脂化學合成物為主，能阻擋雨水滲入垂直的表面。但此成份不能完全避免某些元素所排放的濕氣。

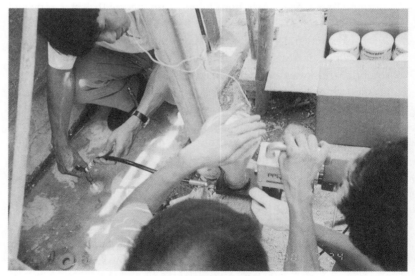

Waterproofing of pipe sleeves through basement wall

Spraying waterproofing membrane

There is a tremendous demand for waterproofing for rooftops. However, instead of the traditional method of renovating rooftops, i.e., opening up all the tiles, insulation layer, and cement sand layer, there are more options nowadays in applying a waterproofing layer on an existing rooftop.

One of the chemicals for the spraying material is called polyurea, which is a two-component elastic coating material for substrate protection from water ingress and corrosion prevention. The protection results in a short time with thickness built up in multiple layers through single-coat application. It has high tensile strength and tear strength. As a renovation material, the cost is low, as no massive demolition is needed. However, it has a rather short shelf life and is more sensitive to moisture and temperature.

Applying sealant at expansion joint

噴塗防水塗料

屋頂防水工程有著巨大的需求，加上屋頂需要經常翻新以杜絕天面漏水，除了傳統的屋頂修補方法，如必須翻開所有的鋪瓦、隔熱層和泥沙層之外，今天還可選用直接噴塗於現有屋頂上的防水塗層。

其中一種噴塗防水材料為聚胺酯，這是由兩種劑料結合而成的軟膜噴塗層，適用於屋頂天面以防止水氣入侵。而且，它只要很短時間就可做到所需的厚度作防水，並具有高韌性及耐磨性。作為翻新天面的防水層，它比起傳統式的做法便宜，不過也有本身的限制，如較短的有效期及對溫差變化的反應。

Expansion joints

Movement joints are provided to allow expansion and contraction of adjacent materials, vibration, wind forces, and settlement of the structure.

For a long horizontal structure, say, a podium supporting residential towers, the structural system is separated to provide the expansion joint at about 100 m intervals. For external walls finished with rendering and tiling works, movement joints separated at the rendering base are provided at about 3 × 3 m horizontally and vertically or at each storey interval, especially for high-rise structures where the risk of falling objects is high. Curtain wall and cladding panels are individually anchored to provide for the expansion joints. For parapet and fence walls, joints at approximately 30 m intervals separating the structure will allow movement. A large area of flooring provided with movement joints can eliminate some of the future maintenance problems. For road works, joints filled with asphalt are provided.

Installation of expansion joint

Nowadays, joineries are more than just functional requirements but more about aesthetics; as such, varieties of proprietary products of expansion joints can be found in the market, especially in Hong Kong. Most residential developments are built with a mega podium mall. Some material matching expansion joints are used in podium malls, for example, the Elements in West Kowloon District.

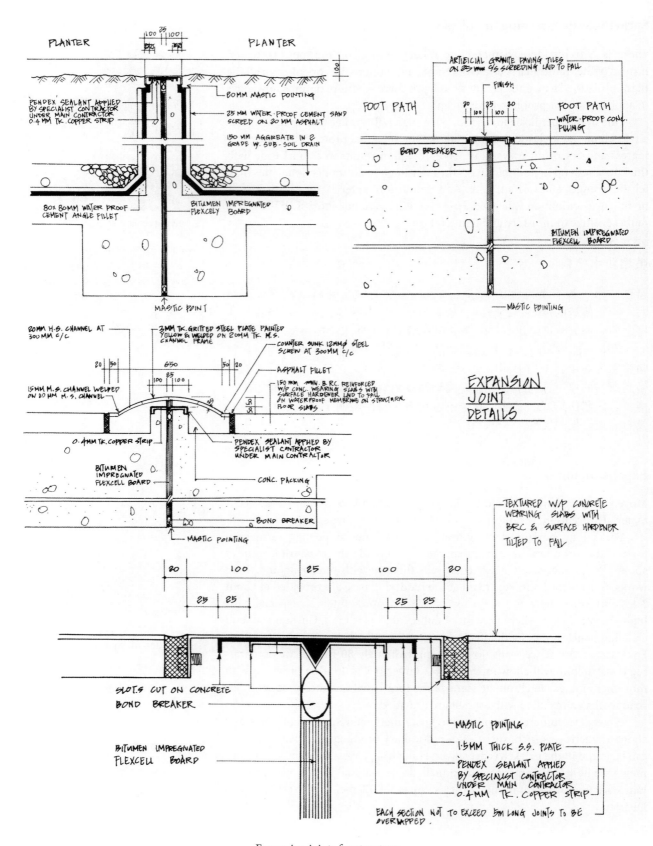

Expansion joints for structure

伸縮縫設計

建築上的接駁縫，容許物料之間因著震盪或結構上的沉陷而形成伸縮。

以一個長距離的橫向結構，如支撐著住宅樓宇的平台來說，須在約100米的距離之間，預留伸縮縫作結構上的設計。外牆表面的磚瓦工程中，亦要設計有約3×3米縱橫交錯的伸縮縫，這點對於高樓建築物尤其重要，因高空跌下的物料會造成更大的危險性。

至於玻璃幕牆外鋪砌鍍面板工程，主要是以獨立的安裝結構組合，因此需要預留個別伸縮縫。建築露台護欄和圍牆時，接駁的結構間要以約30米的距離留伸縮縫位置。而於地面建築中，必須在適當的距離留有可容許的伸縮縫。一般路面都是以瀝青來填塞伸縮縫的。

現今的設計師對於接合處的要求不止是著重功能，而更多是在美觀上。因此，市場上有很多伸縮縫的預製產品。特別是在香港，很多住宅建有大型的平台商場，一些配合材質的伸縮縫應運而生，多用於平台商場的地面，如西九龍的圓方商場。

Finished material matching expansion joints for structure

Sealants

To provide for both movement and weather protection, sealants are used to fill the joints. Building sealants available include oil-extended mastics, oil-based caulks, butyl sealants, latex acrylic sealants, solvent-based acrylic sealants, polysulfides, silicones, polyurethanes, and sealants. The structural sealant can support glazing in a frameless glass construction.

Oil-extended or oil-based caulks are suitable for temporary structures that undergo little or no movement. Life expectancy is two to ten years. Failure occurs by hardening, shrinkage, and cracking. Movement is up to 5% joint width.

Butyls can allow greater elasticity, but there are problems of tackiness and staining, due to the plasticizer. The expected lifetime is 10 to 15 years. Movement is up to ±10% joint width.

Acrylics have good unprimed adhesion and good weathering characteristics but a strong, obnoxious odour. They have a 10- to 15-year life expectancy outdoors and can accommodate movement up to ±12.5% joint width.

Polysulfides are inexpensive and popular and can move up to ±25% joint width. Except for steel, glass, and aluminium, priming is required on the substrates. Application on glass windows is quite common.

Silicone has good thermal stability and good resistance to ultraviolet degradation. An expected lifetime can be over 20 years. Aesthetically, this is a clear sealant which is preferred on some occasions. Movement is ±25% joint width and can be ±50% with very high-performance sealants. Nowadays, a very wide range of silicone is available to suit different applications.

Urethane sealants form another type of high-performance sealant. Most characteristics of urethane sealants lie between polysulfides and silicones, but urethane is more susceptible to problems with dampness in the applied surface. Movement allows ±25% but can go up to ±35% for some urethanes.

封 料

　　封料具有伸縮功能，兼備防水性質的物料。建築用封料包括油性膠黏料、油性堵縫劑、碳合成封劑、聚硫化合物封劑、矽以及特別封料劑；另有結構性封料，用於支撐無框玻璃。

　　油性堵縫劑主要用於暫時性結構，而結構本身應以較少或完全沒有任何活動性質為主。此類物料可具備二至十年的效能。強化、收縮和裂縫會導致物料毀壞。預期活動範圍最寬為封口的 5%。丁基橡膠有較大的伸縮性，但因有增塑劑而較易發黏及染色，預期效能為十至五年，預期活動範圍最寬為封口的 ±10% 左右。

　　塑膠料具有良好的黏附性和防水特質，但含有強烈較難聞的氣味。戶外使用的預期效能為十至十五年，活動範圍最寬為封口的 ±12.5% 左右。

　　聚硫化合物封劑較經濟和普及，活動範圍最寬為封口的 ±25%，除了鋼、玻璃和鋁以外，需要附加底層處理工夫在物質表面。一般用於玻璃為主。

　　矽具備良好的傳熱性能和防止紫外光的破壞，預期效能可以超過二十年。從外觀來看，一般以透明的矽較為廣泛使用。至於活動範圍最寬為封口的 ±25%，而高性能矽更達 ±50%。

　　氨基乙酸甲脂封料是另一種高性能的建築用料，它的特性介乎聚硫化合封料與矽之間，但功能可因著物質表面的濕氣而出現毛病。活動範圍最寬為封口的 ±25% 至 ±35% 不等。

2.8
Carpentry, Joinery, and Ironmongery
木工、細木工、五金

Materials

Timber for carpentry and joinery (more refined and finished woodwork) is generally to be of mature growth, well and properly seasoned, free from wood wasp holes, holes, large loose or dead knots, splits, or other defects that will reduce its strength.

On site, timber is stored in a dry, well-ventilated place, protected from the weather.

Moisture content is calculated as:

$$\frac{\text{Wet (or supplied) Mass} - \text{Dry Mass}}{\text{Dry Mass}} \times 100 = \text{Moisture Content Per Cent}$$

The dry mass is obtained by drying in an oven at a temperature of 103°C ± 2°C until the weight is constant.

The maximum permissible moisture content in timber is as follows:

1. Internal timber for use in air-conditioned space, 12%
2. Internal timber generally, 16%
3. Timber with one side to the exterior of the building and one side to the interior, 18%
4. External timber, 20%

Door design: art nouveau, Artists' Colony, Darmstadt

Artists' Colony, Darmstadt

Columns with bird's-eye maple veneer

Cutting and fixing skirting

Painting skirting

Classification of wood

In any case, structural timber should not exceed 22% moisture content.

Hardwood from broadleaf trees for carpentry or joinery includes teak, san cheong (Kapor), ying muk (Kruen), ash, oak, beech, and maple. The density of hardwood is to be 720 kg per m³ minimum.

Softwood from coniferous trees for carpentry or joinery includes cedar, pine, spruce, and China fir.

Teak from Myanmar or Thailand is widely used in construction works as flooring, doors, and door frames, skirting and cabinets. San cheong is a less expensive material with similar applications.

Canadian maple is a good flooring for squash courts, affording a resilient surface. However, it is highly susceptible to dampness and should be worked and maintained in a conditioned environment.

Ash, oak, and beech are used more in interior decoration than in general construction and give good quality finishes.

Plywood is extensively used in general construction and interior work. Grade 1 veneer has teak or other hardwood faced for a natural finish. Grade 2 veneer has luan faced for painting. The bonding adhesive between plies is resin adhesive classified as moisture and weather resistant. For external use and areas of high humidity, the bonding adhesive is of weather- and boil-proof class.

Laminated plastic sheeting is used in close association with plywood glued by synthetic resin adhesives forming a water-repellent surface which is easy to maintain.

For a natural finish, Grade 2 veneer has luan faced for painting. The bonding adhesive between plies is resin adhesive classified as moisture and weather resistant. For external use and areas of high humidity, the bonding adhesive is of weather- and boil-proof class.

Laminated plastic sheeting is used in close association with plywood glued by synthetic resin adhesives forming a water-repellent surface which is easy to maintain.

Boardwork used in association with carpentry and joinery includes blockboard, hardboard, insulating board, wood chipboard, and fibreglass insulating board.

物 料

用於木工、細木工項目的木材，是來自成熟的、經小心處理後的木料，並且須選取那些沒有細孔、死去細胞紋或其他病症的樹木，才可確保木器本身的強度穩定。

木中的濕度成份可以下列方式計算：

（濕木質量 − 乾木質量）÷ 乾木質量 ×100 ＝ 濕度百分比

計算乾木質量的方法是把木條放在爐中以 103°C ±2°C 的溫度烘乾，取其穩定後的質量。

容許最多濕度的貯量：

(1) 可控制室溫於室內用木材　　　　　　　12%
(2) 普通室內用木材　　　　　　　　　　　16%
(3) 建築物運用（只有單一面外露於空氣）　18%
(4) 外用木材　　　　　　　　　　　　　　20%

在任何情況下，結構用木材的濕度量都不能超過 22%。

由大葉樹而來的實木,可用於木工和細木工項目,當中包括柚木、山樟、英樸、槐木、橡木、水青網、楓木等。硬木的標準密度為每平方米 720 公斤以上。軟木源於不落葉的植物,包括杉木、松木、檜、中國冷杉木。

來自緬甸、泰國的柚木,較常用於地板、木門、門框、木腳線和衣櫃的工程,但較經濟的山樟木亦是常用的木料之一。

加拿大的楓木,富有彈性,普遍用於壁球場的木板地面。但木材本身較容易受潮濕天氣所影響,因此運用時須特別小心及控制,在固定室溫下進行工程較為理想。

槐木、橡木和樺木較常用於室內裝飾工程,效果較佳且美觀。

夾板是最常用於工程裝飾項目的木材。甲級表面板含有柚木或其他實木的天然表面成份;乙級表面板須有凹凸表面作塗漆工作。如用於戶外,因常暴露於潮濕的天氣下,夾板間黏膠成份中的直脂須選擇為能抵禦天氣和防暴曬的類別。

由多層夾板製成的膠板,是以人造植脂黏貼成防水的表面,既有防水作用又較容易打理。

其他的木工材料包括硬木板、防熱板模、碎木屑板模、玻璃纖維防熱板。

Dipping in hot bitumen

Cutting machine

Flooring

Boarded or strip flooring is of teak or other hardwood with a finished thickness of 20 mm minimum. Width for solid teak is 75 mm and 100 mm; that for solid maple is 50 mm. Length varies. The floorboards are tongued and grooved on edges and heading joints. Battens of hardwood 50 × 40 mm pretreated with wood preservative are laid at 350 mm centres on a concrete bed level with 1:3 cement mortar. Spaces between battens are infilled with lightweight concrete. A coat of bitumen or rubber latex emulsion at the rate of 1 litre per m² is applied over the entire surface. After it dries a second coat at 0.67 litre per m² is applied. The floorboards are then secret-nailed onto the battens.

As a common form of domestic flooring, woodblock is of teak or other hardwood. Block size is 300 × 50 mm with 20 mm minimum finished thickness. Since the colour of natural wood varies, it is good practice to sort and match for colour before installation. Blocks are tongued and grooved at the sides and ends. Fixing is done by dipping the underside of the blocks in heated bitumen and laying on the screed coated in similar bitumen. Laying is done in a basket or a herringbone pattern (better quality) with straight borders minimum two blocks wide. At the perimeter of flooring, a 5 mm expansion gap is filled with cork strip or foam rubber strip. Sanding is then done by using an electric surfacing machine with various grades of abrasive paper to obtain a smooth surface to be coated with sealer or polish after filling in the gaps with sawdust.

A lower-standard of finish is parquet flooring of hardwood with block size 120 × 25 × 8 mm thick. Fixing is done by dipping the undersides in Ucar™ Latex or equal adhesive and laying on a screeded bed coated with similar adhesive. After sanding and filling in, the even surface is ready for wax polishing.

Sawdust to fill gaps

Sanding machine

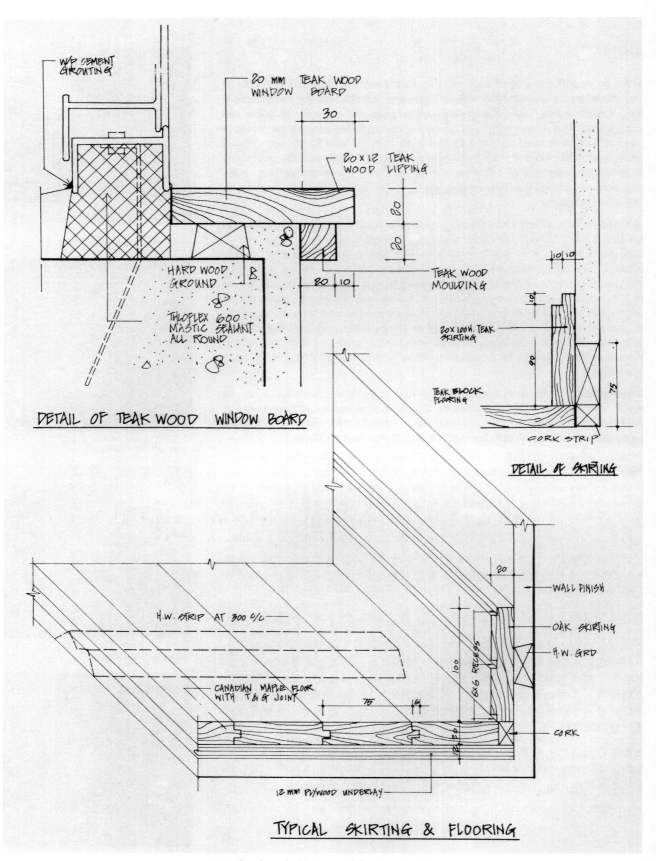

Window board, skirting, and flooring details

地 板

木板塊、木條地板來自柚木或其他實木，厚度最少有 20 毫米。實心柚木地板的闊度由 75 至 100 毫米，楓木由 50 至 100 毫米闊，長度不等。木板塊均在邊旁留有齒頭和凹陷的地方作完整的接駁口。先由 50×40 毫米的小木塊，經過保護膜的處理，放在 1∶3 的水泥沙混凝土面上，距離不能多過 350 毫米。在木塊間澆輕質混凝土，隨後再以每平方米 1 公升的密度加上一層瀝青或橡膠料，乾涸後再以每平方米 0.67 公升的密度加第二層的瀝青面，最後才可以裝嵌木條地板。

　　住宅常見的地板是柚木或其他實木類。條型的度量為 300×50 毫米，物料經施工完成的厚度最少為 20 毫米。由於天然木材的色澤較為參差，最理想的做法是在施工前，揀選顏色較為相近的，才開始鋪砌。條型木塊旁邊開齒狀口和凹陷的地方，能使木板之間互相扣緊。鋪砌前，先把基礎底層的砂漿面塗上瀝青，再以木條底面塗上熱瀝青油。鋪砌時，以篩紋或魚骨式鋪排較好，地板圍邊以兩塊木條為組合處理。在地板的周界，配以 5 毫米的水松條或發泡膠條作為伸縮縫。隨後用打磨器把地面打沙、磨平滑，方便日後進行蠟面等工作。

　　較為廉價的普通地板是以實木條 120×25×8 毫米所組成的方塊地板料。鋪砌時，先將地板底面塗上 Ucar Latex 或類似的膠漿，再鋪在已做好的泥水地面，經打磨和蠟面處理後便告完成。市面上不難找到已完成臘面處理的地板塊。

Engineered wood flooring

Engineered wood, also called composite wood, is commonly used in domestic flooring and ranges in size and appearance. It is composed of several layers of wood. Hardwood is the surface layer. Underneath is plywood in different grain directions so that the expansion of the materials compensate each other for better performance. This kind of wood has a tongue-and-groove system for easier installation. The hardwood surface gives a natural appearance and allows polishing as normal hardwood flooring.

複合木地板

複合木地板常見用於住宅地板。這種材料有不同程度的大小和外觀，主要由幾層的木材組成。外層是實木，中間是不同紋理方向的夾板。因此，材料的擴張以互相補償方式得到更好的制衡。這種材料有易於安裝的榫槽，實木表層提供自然的外觀，且可與正常實木地板一樣打磨及拋光。

Laminated wood flooring

Another popular wood flooring nowadays is called laminated wood. Different from engineered wood, laminated wood uses high-density fibreboard with a high-pressure laminated surface that provides good impact resistance to the floor. Thanks to technology, the appearance of laminated wood is almost the same as that of real wood and texture, but its high resistance to water and durability is outstanding. Moreover, as it is not real wood, more varieties of colour and texture can be made.

纖維木地板

另一種流行的地板材料是纖維複合地板。與複合木地板不同的是，纖維木地板使用高密度的木屑、膠水，以高壓表層得到有效的地板耐磨力。在科技配合下，這種纖維木幾乎與真實木材的紋理一樣，而且能高效防水且耐用性更久。此外，因為它不是真實木材，可以人工做成更多種類的紋理和顏色，可以說比起真木有更多的花樣。

Doors

A common way to construct solid hardwood doors is with panelled doors. The thickness of the door is 40 mm minimum with 100 mm wide styles, top rail, and muntins, and 200 mm wide middle and bottom rails. Panels for glass can be installed.

Flush doors are a common method of door construction in Hong Kong. Stiles and rails generally are 75 mm wide. For doors exceeding 900 mm wide or 2000 mm high, stiles are increased to 100 mm wide. Hollow-core flush doors are infilled with 25 mm horizontal battens at 125 mm centres or 25 mm egg-crate battens at 125 mm centres for wider and stronger doors. Solid-core flush doors are infilled with 25 mm vertical battens tightly cramped together or with 50 mm horizontal battens. Covering both sides of solid- or hollow-core doors can be 5 mm plywood for painting, 5 mm hardwood or teak veneered plywood for natural finish, or laminated plastic bonded to 5 mm plywood. Hardwood or teak lipping of 12 mm is glued to all edges.

Ledged doors, ledged and braced doors, as well as framed, ledged, and braced doors forming another type of timber door are not widely used.

Door frames of solid hardwood are fixed to concrete floors with dowels inserted not less than 75 mm into the bottom of posts and fit in mortar. Fixing to walls is by mild steel bolts (2 nos. on each side) for concrete and holdfasts (3 nos. on each side) for brickwork or blockwork. A 150 mm deep precast concrete lintel extending 225 mm on both ends is fixed at the top.

Architraves are fixed in one length between angles with mitre joints at the angle. Generally, fire protection with a half hour fire-rating period can be achieved by a minimum 44 mm thick solid timber door. The door frame will have a timber doorstop continuous round the sides and the head of the frame and a dimension parallel to the face of the door of not less than 25 mm and at right angles to the face of the door of not less than 36 mm. One-hour fire-check flush doors are available as approved proprietary fire doors with a thickness of 54 mm. All fireproof doors including the door frame should be tested in accordance with or assessed against BS 476: Parts 20 to 24:1987 and certified as being capable of resisting the action of fire for the specified periods. Cupboard or closet doors can be constructed as hollow-core doors or with plywood or blockboard lipped on all edges and faced with laminated plastic or hardwood veneer finish.

門

格板門是最常見的實木門。門的厚度最少為 40 毫米，並有 100 毫米闊的門挺、上冒部份；200 毫米闊條擺放在鎖冒及下冒。亦可安裝玻璃面板代替門芯板。

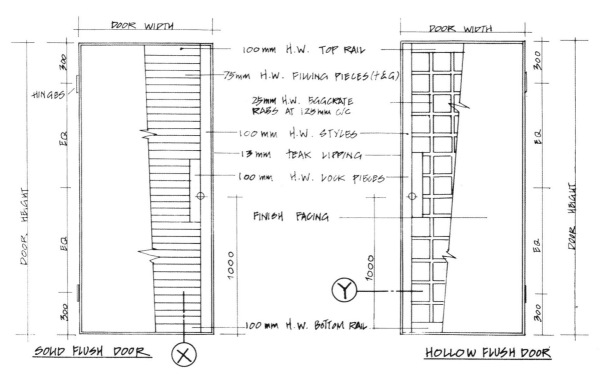

Solid and hollow door construction

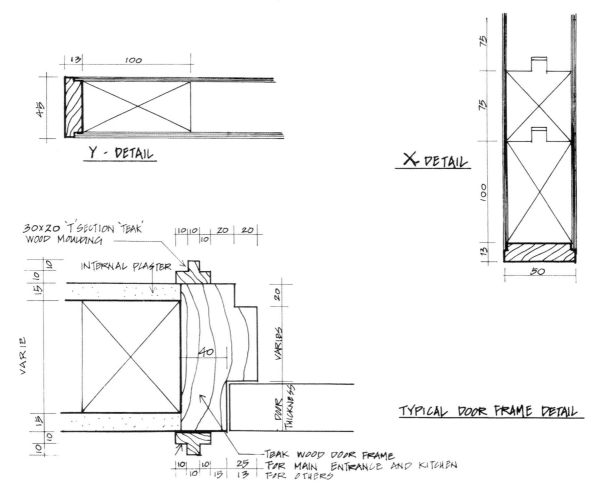

Door frame details

Door frame installation

Solid core flush door

Flush door with louvre

香港常用的平面門，以 75 毫米闊的木條為豎框和中間的片塊。當門的尺寸超過 100 毫米闊或 2,000 毫米高時，豎框需增加至 100 毫米闊。空心的平面門由 25 毫米的橫向小木條，自中至中的距離 125 毫米；或者是 25 毫米的蛋格小木條，同樣的 125 毫米距離，用在比較闊和承力較強的門。實心的平面門由 25 毫米直向的小木條或 50 毫米橫向的小木條，緊密地連結在一起。覆蓋在實心或空心門的物料包括以油漆粉飾的夾板、5 毫米的實木或柚木表面板；或再加防火膠板。最後用 12 毫米的柚木或其他實木作為上下封邊。

實木門的門框穩妥釘於混凝土地面上，且用綴縫釘打入在最少 75 毫米深的混凝土地面上，隨後再以砂漿封好。連接於垂直牆身時，混凝土邊以每邊兩個鋼鐵釘縫緊；要是牆身為磚塊，則以每個邊三個繫鋼釘鑲穩。在門頭頂，安裝 150 毫米深的混凝土楣梁，每邊伸展多 225 毫米。封口線則以斜榫接方式安裝。

消防用的半小時防火門通常須有最少 44 毫米厚的實木，同時，木門框要有門檔板，延續於門邊及頂，平衡於門面最小有 25 毫米厚，直角於門面最小有 36 毫米闊。至於特製的一小時防火門則通常需要有 54 毫米厚。以上所述各類防火門連門框，都需要通過英國標準之測試並取得合格證明書。衣帽間的門可由夾板和各式面板組合成的物料，旁邊四周再鑲以膠條或實木。

Ironmongery

Ironmongery is selected by reference to the function of the door fixtures, particular style, colour, and finish. Specification is by brand name, model number, and catalogue reference. The keying system and door thickness for locks must be stated.

There is a wide range of locks and latches available for selection in the local market and manufactured in Japan, Germany, the UK, the US, Australia, and other countries. A lock is activated by means of a key, whereas a latch is operated by a lever or a bar. The exact type of lock should be considered in relation to the use of the door, the detailed dimension for door thickness, and the rebate on the door frame, so that the knob can be easily operated by hand after installation.

Materials and finishes include stainless steel, electroplated coatings of nickel and chromium, electroplated coatings of cadmium and zinc on iron or steel, and phosphate treatment of iron and steel. Standards should conform to relevant BS. Screws used in association with ironmongery are stainless steel, brass, or aluminium with counter sunk heads.

Hinges are classified by their function, length of flap, material used, and perhaps by method of manufacture. The correct number and type of hinge are specified to suit the correct operation of the door, window, or gate. A steel butt hinge is the double flap butt hinge, rising butt hinge, parliament hinge, tee hinge (strap hinge), spring hinge, bank and hook, etc.

Overhead door closers keep a door self-closing, which is required for fire-rating doors in kitchens and smoking areas. The closing speed of the closer should be made adjustable. A concealed door closer is not visible when the door is closed.

Other items of ironmongery are magnifying viewers, barrel bolts, door-stops, indicating bolts, flush bolts, pull handles, push plates, cupboard catches, drawer pulls, shower curtain rods, hooks, and so on. Details and designs can be referred to in numerous manufacturers' catalogues.

五金

五金的選擇，取決於門的作用、類別、色澤和表面的處理。招標文件中通常須列明它的類別、型號和供應商，而門的厚度和門鎖系統決定了五金的選擇。

在香港，五金的選擇很多，有來自日本、德國、美國、澳洲和其他多個國家。門鎖運作的原理，主要是以鎖匙開動鎖中的杠杆。至於門鎖位置的設計，則要考慮門的功用，另需要顧及一般使用者的方便。

門鎖物料的選擇分別有不銹鋼、電鍍鎳或電鍍鋁合金面、鋁合金、鐵或鋼面上電鍍鎘和鋅、以硫化磷處理的鐵或鋼面。品質以英國標準為依歸。所有用於五金上的螺絲釘以不銹鋼、黃銅或鋁等合成，螺絲帽要以平頭為標準。

鉸鏈分類以其本身的功用、長度、物料，甚至製造方法來決定。正確的型號和鉸鏈類別須列明於標書中。鋼鉸鏈是較為常用的種類，其他選擇則包括雙明合鉸鏈、旋升鉸鏈、門邊鉸鏈、平型鉸鏈、彈簧鉸鏈等等。

門頭上閉門器具備自動開關門功能，適用於廚房、防火大堂的防火門上。自動關閉的時間和速度可以調較來適合不同需要。如需美觀的設計，可以選擇暗藏的閉門器。

五金項目還包括門前魚眼鏡、筒形插鎖、門碰頭、顯示螺栓、平面栓、拉手板、推手板、抽屜手挽、浴室簾桿等。供應商可收提供不同類型的款式設計。

Carpenter's tool

List of common locks
常見鎖匙

Lock type	Series	Remarks
Cylindrical locks	Heavy duty	• High level of security and durability • Wide range of applications
	Hotel lock	• Outside knob is always rigid • With guest key, master key, and emergency key
	Standard duty	• Convenient and easy to operate • Wide range of applications
	Light duty	• For lightweight doors such as bathroom, toilet, passage
Mortise lever handle locks	Lever handle lock	• Effective for security • Wide variety of designs • For entrance, classroom, office, conference
	Lever handle latch	• Easy operation • For passage
Mortise locks	Heavy duty	• Effective for security • For entrance, classroom, office, storeroom
	Standard duty	• Effective for security • Smooth and easy operation • For entrance, classroom
Rim locks	Lever handle type, knob type, grip handle type	• Effective for security • Strong, durable design • For apartment, condominium entrances
	Rim deadlock	• Economic and compact • For auxiliary locks, inspection doors, swinging doors, and sliding doors • For narrow lock stiles of doors
Deadlock	Mortise deadlock, night latch	• For doors which do not need knobs • Inspection doors, auxiliary locks, storerooms • Alternative design for glass doors
Sliding door locks	Single sliding door lock Double sliding door lock	• Strong hook bolt lock • For sliding doors
Flush cup handle locks	Mortise lock, integral lock	• For storerooms, gymnasiums, fireproof doors
Emergency exit door locks	Mortise deadlock Mortise electric lock	• For emergency exit doors • Access to a street can be fitted with panic bolts
Decorative Locks	Mortise locks, rim lock, cylindrical lock, deadlock	• Deluxe requirement • For entrances
Electric Locks for a building's security system	Mortise electric lock	• Locked or unlocked by flow of current • For building entrance control, interlock doors, etc.
	Card lock system	• Security system for entrances, hotel guest rooms, secret rooms • Operated by a magnetic card

2.9
Staircases, Steps, and Handrails
梯級及欄杆扶手

General design basis

Staircases and steps provide movement of people between different levels and are a means of escape in case of fire. As a detail design, these very often represent the first point of architecture that people physically experience. The texture, colour, material, and dimension of steps and railings form intimate relationships with users.

As the means of escape in case of fire, the design of a staircase is governed by the Building (Planning) Regulations and the Code of Practice on Provision of Means of Escape in Case of Fire and Allied Requirements.

However, from the developer's point, fire staircases are 'non-usable' floor area and should be designed to occupy the minimum floor area. As a result, the 'scissor' staircase formed by two adjacent flights of 16 risers is a common feature in Hong Kong buildings.

設計原則

梯級的基本功用是讓人從某一個高度走到另一個高度，以及在遇火災時作逃生之用。作為細部設計方面，梯級代表建築物與人們的第一個接觸。梯級及扶手的質感、顏色、物料和尺寸都會對使用者造成影響。

走火逃生樓梯的設計須由建築事務監督批准，並合乎《建築物（規劃）規例》和《1996年提供火警逃生途徑守則》。

但從發展商的角度來看，由於逃生樓梯的設計非真正實用的面積，一般來說，他們都只會從經濟實用的方向去考慮。「鉸剪樓梯」的設計正好是把樓梯佔用的空間減至最少，通常以十六級為一段，這亦是香港建築設計的特色之一。

Material of staircases

Every building exceeding two storeys has the main staircase constructed with a fire-resisting period of not less than one hour (now calculated in minutes, that is, 60 minutes) or the period required for each element of construction of the building or compartment in which the staircase is situated (whichever is greater).

Reinforced concrete staircases are non-combustible, strong, and hard-wearing. Usually casted in situ, concrete is the most common material used. Steel staircases are constructed in some instances where fire-rating is not required and a lightweight construction as well as an elegant outlook is appreciated. Timber staircases are used more in temporary work.

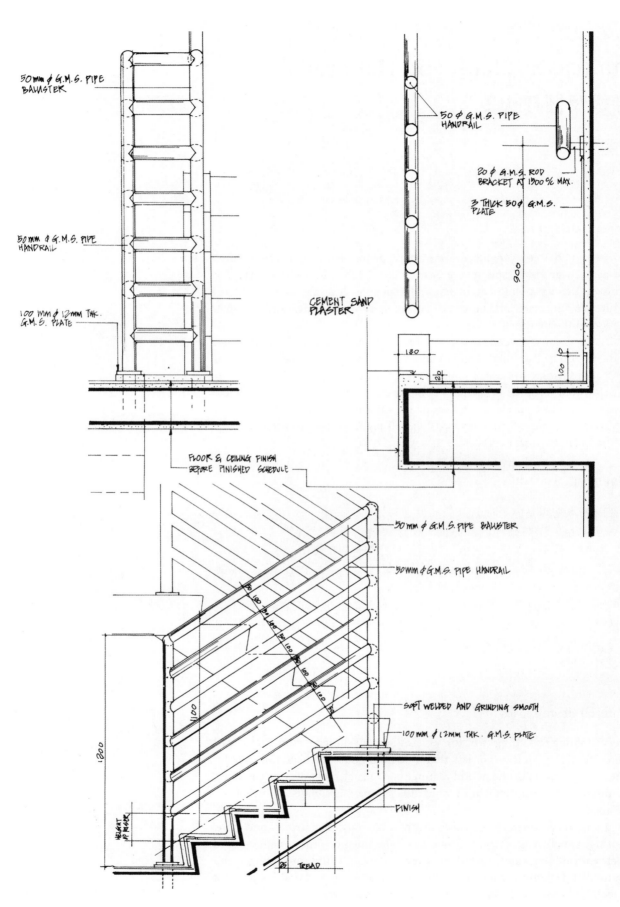

Staircase construction details

Finishing to a reinforced concrete staircase can be plain cement screed, mosaic, or unglazed tiles with nosing tile. For grand staircases such as in hotels, a marble finish is often used. Consideration of the non-slip surface of steps should be made, e.g., an addition of carborundum strips, for safety reasons.

樓梯材料

凡超過兩層樓的建築物都須建一條防火樓梯,防火標準以一小時(現時量度單位為分鐘計,即 60 分鐘)防火時間或用樓宇空間體積的防火時間(最大的系數)。

混凝土樓梯是非助燃性、堅固和耐用的,建造方法通常是在地盤實地做紮鐵釘板,然後澆灌混凝土。鋼鐵樓梯常用於不須要防火樓梯要求的地方。木樓梯則只用於臨時的地方。

樓梯表面的物料包括水泥砂漿、紙皮石、啞面瓷磚和級嘴瓷磚。而用於較為華麗的地方如酒店大堂等,都會在混凝土樓梯面鋪上雲石。在選擇物料時,要考慮用不會滑溜的邊瓦,或者是用鋼砂條,作為防滑的設計。

Fire escape staircase

Requirement for staircases and steps

The staircase is to have a clear height of not less than 2 m and a clear width of not less than 900 mm. Treads are not less than 225 mm wide, and risers are not more than 175 mm high. For more than 16 steps in a flight, a landing must be introduced; neither are there fewer than two risers in a flight. For schools, treads are not less than 250 mm wide, and risers are not more than 150 mm or less than 75 mm high.

Landings must be provided at the top and the bottom of each flight not less in width and length than the staircase. It is important that the swing of any exit door will not reduce the effective width of the landings.

The 'scissor' staircase

樓梯設計的要求

法例註明,樓梯淨高最少要 2 米、淨闊 900 毫米。踏板的最小闊度為 225 毫米而板高最大為 175 毫米。當一段樓梯超過 16 級時,要加上平台才開始繼續第二段的樓梯級。樓梯級不能少於兩級。至於學校的梯級,踏板要調整為最小闊度 250 毫米、板高則在 75 至 150 毫米之間。

樓梯著陸間的闊度要以樓梯的闊度為基本。重要的是逃生門門掩不能阻礙樓梯著陸間的闊度。

Steps at the Bond Centre

Handrails

Handrails for staircases are required on each side of the stairway, except for stairs less than 1,050 mm wide; then a single handrail at one side of the stairway is sufficient. In the 'Design Manual for Barrier Free Access 2008', there are new requirements for handrails, with more detailed descriptions of handrail sizes, strengths, and shapes illustrated. Braille and tactile information on directional arrows and floor numbers have to be shown at the ends of handrails. A turn-down or return fully to end post or wall face is needed.

Steps at the Hong Kong Park

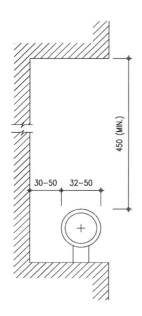

The handrail must be located at a height between 850 mm and 1,000 mm. The projection onto the staircase for a handrail should not exceed 90 mm.

For purely fire staircases, handrails can be constructed of PVC-coated cast iron bars or tubular galvanized steel. For better appearance, stainless steel and timber handrails are possible.

扶手設計

除了闊度不超於 1,050 毫米的樓梯外，一般樓梯的兩旁須設有扶手。扶手高度須有 850 至 1,000 毫米，距離牆身 90 毫米。設計逃生防火樓梯時，要以非助燃性的鐵或不銹鋼作扶手材料。

在《設計手冊：暢通無阻的通道 2008》中有一些關於扶手的規定，包括扶手尺寸、力學和形狀的細節描述。每層樓梯的結束位置，必須設有方向箭咀及樓層號碼的觸覺點字。扶手末端須轉下收入平台面或完全轉後收入末端柱子或牆身。

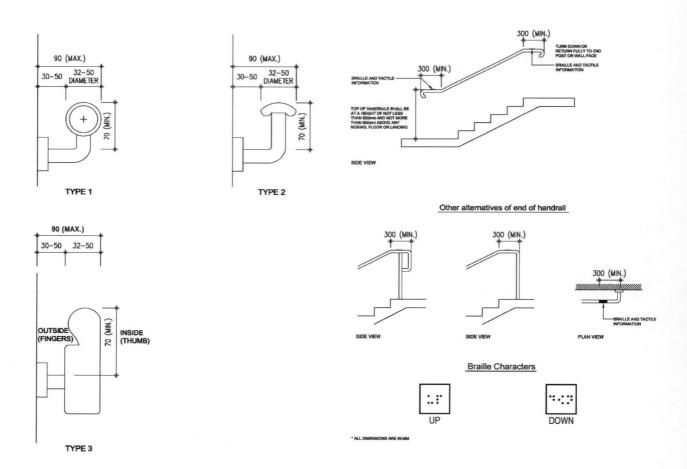

Extracted details from the 'Design Manual for Barrier Free Access 2008'

Steps at the Hong Kong Cultural Centre, Tsim Sha Tsui

Staircase at the Bank of China

2.10
Metal Windows and Doors
金屬門窗

Windows at Laguna City

Window of a traditional-style house in Germany

Lighting and ventilation

Windows and glazed doors function to provide natural lighting and ventilation, which are required by the Building Authority for rooms used as an office, a kitchen, or for habitation, and rooms containing a soil fitment or waste fitment.

For rooms used as habitation, office, or kitchen, the aggregate superficial area of glass in the windows is not less than one-tenth of the floor area of the room, and the openable windows should be equal to one-sixteenth of the floor area. Details of such prescribed window requirements can be applied by using mechanical ventilation and artificial lighting, which is usually applicable to office buildings and hotels.

照明與通風

建築事務監督指定凡擬作辦公室或廚房或居住用途及具備廁器的房間，都必須配備天然照明和通風功能的窗或玻璃門。

《建築物（規劃）規例》註明，擬作居住用途或辦公室或廚房的房間，總玻璃窗透光的面積須不少於房間可用面積的十分之一；當中至少要有十六分之一為可開啟的窗面面積。有些用作辦公室或酒店等的建築，可申請豁免，但仍必須向有關當局顯示建築物設計具備人工通風和照明設施。

Materials

Aluminium for windows, doors, and louvres is manufactured as extruded aluminium alloy. Aluminium sections used in Hong Kong are available from China, the US, Japan, Australia, and many other countries. Aluminium sections have to have a minimum wall thickness of 1.6 mm and be dovetail grooved for weatherstripping. Finishing on aluminium can be clear anodized, colour anodized with average 25 μm thick, or have a proprietary hard coat anodic finish with hardwearing qualities. Spray-applied coatings are possible and range from the most durable premium fluoropolymer coatings through the mid-range modified acrylics to polyesters and thermosetting acrylic enamels. A wide range of colours is available for the latter coatings, which can be durable and long-lasting.

Steel for windows and doors is universal mild steel-rolled medium (32 mm) or large (38 mm), grooved for weatherstrip and free from rolling defects. All manufacturer's fabrication holes are sealed by welding before site delivery. The steel is hot-dip galvanized and the cut ends treated with two coats of metallic zinc-rich priming paint.

物 料

鋁金屬的窗、門、百頁扇門等由鋁合成的物料製造。鋁料除了來自中國外，也有來自美國、日本、澳洲等地方。鋁質的窗框切面須有 1.6 毫米的厚度以及鳩尾槽作密封條。鋁金屬的表面處理，包括厚身的氧化層，電鍍顏色有不同選擇而平均有 25 微米的厚度。除了一般常用堅固的氧化表面處理方法外，還有較耐用的氧化聚合物、中級程度的改良塑膠、熱固丙烯搪瓷等。另塑膠性抽瓷具備多款不同的顏色選擇，且很耐用。

用於窗、門的鋼由軟鋼製成；槽的尺寸分為中級（32 毫米）或高級（38 毫米），為的是防水和碾壓時所出現的問題。供應商必須在門窗運送到地盤之前，把一切裝嵌用的孔洞焊接妥當。鋼是用熱浸鍍鋅，再在剪口塗上兩層金屬鋅化的油漆。

Aluminium windows and doors

Having a neat appearance and being nearly maintenance free, aluminium windows represent the most common type of window used mainly in residential developments in Hong Kong.

As extruded sections vary from different manufacturers, the architect's drawings only indicate the window design in outline and size of structural openings; specifications are performance-type requirement only. The contractor has to prepare his or her own shop drawings, which should solve all problems of water-tightness, thermal or structural movement, pressure equalization, fixture and anchorage, stability, and moisture disposal. Structural calculation based on the wind load at the particular location also has to be prepared by the contractor. Samples of aluminium section, glazing, and sealant materials should also be submitted to the architect. To avoid potential problems of liability, the grouting around the windows must be included in the window contract.

Window frames are mechanically and mitre jointed to develop the full strength of members and to provide a neat weather-tight joint. Drainage holes are provided at bottom members. Fixing of the frame is by galvanized steel fixing lugs spaced at 300 mm centres. Water bars of galvanized steel may be provided for the whole window width. Grouting is waterproof cement mortar applied all around the windows where temporary wooden block sitters are removed. Glass is preferably fixed as internal glazing with a silicone-based or polysulphide sealant. If an additional structure is required, galvanized steel bars can be installed inside the aluminium sections.

Double-glazing aluminium windows or double windows can afford good sound insulation for environmental protection. However, the failure of double glazing is usually due to bad sealing at the edges. Also, for places where the temperature difference is high, moisture can easily be trapped inside the air gap and cause fog on the surface of the glass. Sliding aluminium doors offer good vision but are usually not as weatherproof, especially in typhoon conditions.

To prevent moisture from getting into the interior of the building, there is a good method practised by the Japanese construction industry. Extruded covers are used to cover up intersections between windows and window sills. (See illustration.)

Various window designs

鋁質窗與門

由於鋁窗外觀較為整潔及容易清理維修，故此，香港住宅樓宇一般都會採用。

同時，因為每個供應商都有各自的擠壓型模，所以建築師必須設計窗的基本外貌、結構的實際尺寸等。至於規範說明，則以窗的應用測試結果為準則。承建商方面，必須根據建築師所列明的規範設計畫則，預備施工方案及樣板給建築師批核，當中包括如何防水、冷縮熱脹的移動、結構移動、壓力平衡、裝置固定等細部圖樣、穩固、潮濕和排放水等等各類主要項目；還有，必須提供結構的設計及計算資料，計算的準則以某幾個位置的風力為著點。而鋁窗的切面框、玻璃、封料等資料亦須呈交建築師批核。為了避免可能發生的責任問題，封窗的封料部份，可歸納於窗項目的合同。

窗框以機械安裝方法和斜切的駁口位置，來保持框料的穩固性和防水性能。排水孔洞設計在窗框底部。接連窗框的浸鉛鐵片擺放於 300 毫米的距離。鉛鐵防水帶圍繞窗的四周安裝。封料由防水泥砂漿加在窗的四周，臨時木條亦要完全移走。安裝玻璃時，以矽或多硫化物封料裝好，加強的結構以浸鉛鐵片裝置於鋁框內。

雙重玻璃鋁窗或重扇窗，具有很好的隔音效果。然而，這類窗出現問題的地方，通常是框邊未能完全密封；而且在溫差較大的地方，水份很容易進入空氣間隔層，造成中間玻璃表面起霧。趟軌鋁窗能提供良好的視野，但颱風來時卻未必能完全遮風擋雨。

日本建造業有一個更好的方法防止水氣進入室內，就是在鋁窗外框包一塊鋁質覆蓋板，蓋住牆身與窗台，如圖所示。

Steel windows and doors

Steel windows were commonly used for older buildings but, due to the present popularity of aluminium, they are now used in areas such as lightwells, service rooms, and low-cost housing.

The general method of installation is similar to that for aluminium. Weatherstripping of chloroprene rubber or polyvinyl chloride is also available for fixing into the dovetail groove, providing continuous contact between open casement and fixed frame. Fly screens can be specified and located on the inside of the windows.

鋼窗及閘

鋼窗只能在較舊的建築物中找到，而許多翻新的樓宇，都已經採用鋁質窗框了。

鋼窗的一般裝置方法與鋁窗相似。氧化膠或聚合氧的防水帶可安置於鳩尾縫間，形成可開動窗與固定窗框間的一道連接。防蠅網可以擺放在窗的內部。

Hardware

Hardware selected for the windows is based on the operation of the window, whether it is top hung, side hung, centre hung, or sliding.

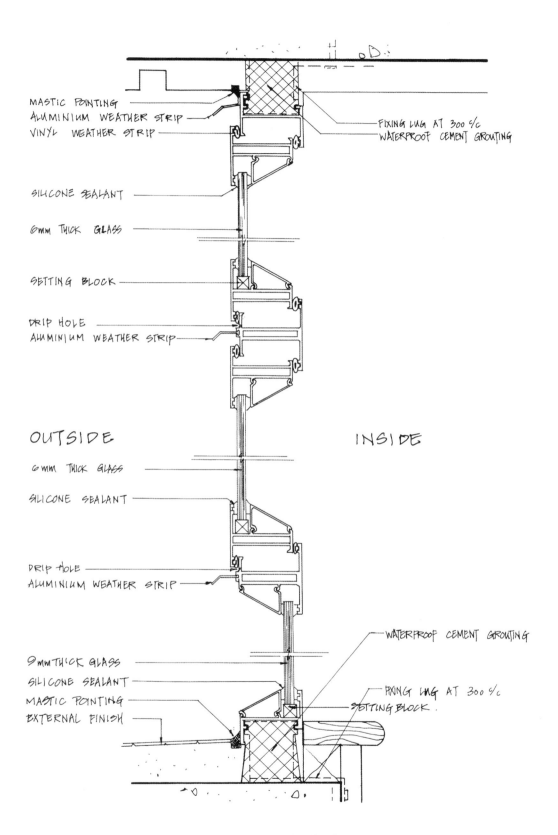

Aluminium window details

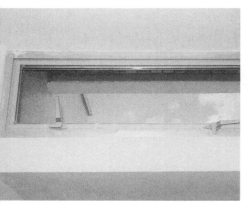

Fanlight

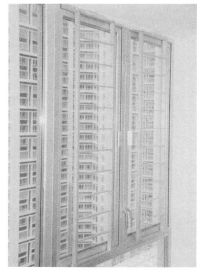

Aluminium window

A stainless steel door in the ground floor lobby of a residential development

Aluminium windows are usually of non-magnetic stainless steel, including:

1. concealed bar parallel link arm hinges in a duel-action locking handle for horizontal slide-projecting window (side hung);
2. concealed bar parallel link arm in a locking handle or typhoon bolts (for fan light) for vertical slide-projecting window (top hung);
3. heavy-duty friction pivot hinges with a duel-action locking handle for horizontal and vertical centre pivot window;
4. floor-spring pivot hinge and mortise deadlock for side-hung door;
5. overhead roller and track, guide roller together with handles, flush bolts, and locking devices for sliding window and door.

Nowadays, a multi-lock system is usually applied in new residential buildings. Compared to a single-lock system, the multi-lock system is operated by a handle. Windows can be operated with one hand. For ordinary domestic use, three pairs of locks (top, middle, and bottom) are located at the window panel and the frame. The advantage is that the sealing action between the window panel and the frame is enhanced so that the performance in deformation and insulation of the window are improved. This system is also used in curtain wall systems with openable windows. The number of locks used depends on the size of the opening.

五 金

選擇玻璃窗的五金時，必須依據窗的功能，如吊頭窗、邊開窗、中掛窗、滑動窗等。

至於鋁窗物料，多用以下無磁性不銹鋼：

(1) 暗藏片平衡連接鉸鏈，要放在雙重功能的手鉸上，用於橫向伸展的滑動（側掛）窗。
(2) 暗藏片平衡連接鉸鏈，放在鎖鉸或防風螺栓（高位的窗），用於垂直的滑動（頂掛）窗。
(3) 重型磨擦式連桿絞鏈，兼具雙重功能的鎖鉸，適用於橫向和直向搖窗。
(4) 地面的彈簧連桿絞鏈和插鎖，用於（側掛）門。
(5) 高架滑輪和軌，輔助的滑輪與把手，的平頭螺栓和鎖，適用於滑動的門與窗。

現今多點鎖系統已經廣泛用於新型住宅。與單一鎖系統不同的是，多點鎖系統由一個手柄操作並控制兩個以上的鎖點，而且窗戶可以用單手操作。應用於普通住宅，窗板和窗框上有三對鎖（上、中、下），其優點是增強窗板和窗框的密封程度，因此窗戶減少了變形，但隔熱力卻增強。這種系統同樣也用於可開窗的幕牆系統，鎖的數量則取決於窗的大小。

2.11

Glazing, Curtain Wall, and Cladding
裝配玻璃、玻璃幕牆與外牆板模

Glass

As one of the oldest human-made products, the discovery of glass was in the Bronze Age. Today, glass is one of the most important building materials controlling people's environment and taking on artistic, beautiful expressions.

Manufactured from soda, lime, silica, and other minor ingredients such as magnesia and alumina, glass is a suitable material for glazing. The raw materials are heated at 1490–1550°C to produce a molten state of glass. Formation of glass sheets is by various processes of drawing, floating, or rolling.

The composition of glass must have adequate durability as per BS 952: Part 1:1978:

- silica and alumina together not less than 71%
- alkalis (Na_2O; K_2O) not greater than 15%
- lime and magnesia together not less than 10%

Rose windows at the Cathedral of Ulm, Germany

Float glass of 6 mm with a nearly undistorted vision and reflection is widely used in windows of local residential projects. For clear glass, light transmission is 85%. For curtain walls, high-performance heat reflective glass 12 mm thick is commonly used with light transmission as low as 10% to 20%. Sheet glass, because of its distortion, is not often specified in buildings.

For safety reasons, laminated and tempered glass are used. Tempered glass is three to five times stronger than ordinary glass of the same thickness and will shatter into small, blunt, drop-like particles when broken, thus minimizing possible injury. In contrast, due to the tough interlayer, laminated glass will not drop out if broken. As tempered glass cannot be cut after tempering, exact sizes have to be pre-ordered, which may mean a long wait time for delivery. Now, local tempering is possible, e.g., Saint Gobain Tempering Factory Limited produces tempered glass up to 2600 × 3400 × 19 mm thick.

Glazing at Bauhaus, Dessau

One of the major design considerations in glass technology is thermal crack. This is caused by extensive sunshine on external glazing, causing thermal expansion at the glass exposed to light, but the glass edge shaded by window frames is not expanded; thus, the thermal differential produces tensile stress along the edge of the glass. If the strength of the glass fills to take up such stress, thermal cracks may occur in the window. Large pieces of glass, heat-absorbing glass, and wired glass are particularly vulnerable. Prevention measures include the following:

1. Thermally insulate the glass from the metal sash. For a curtain wall, clearance between the glass and the sash should be over 4 mm.
2. Avoid building up heat at the back of the glass window, such as placing thick curtains or installing reflective film inside the glass surface.

3. Install a good silicone elastic sealant to resist moisture collection and to provide a bit of movement.

Wind-pressure-resisting strength for a piece of glass depends on the thickness of the glass, the type of glass, how the edges of the glass are fixed, and the dimension of the particular glass. The actual wind pressure on a locality depends on the level of the area elevated from the sea, e.g., at sea level 1.2 kPa; at the Peak 4.3 kPa. On a tall building, local wind pressure varies and can be ascertained by using a model for a wind tunnel test.

There is a wide range of colours for glass, depending on the thickness, the colour of the tint, and the colour of the coating, producing colours like silver, bronze, grey, green, blue, gold, pink, etc. In the selection of the right type of glass, large samples of actual colour and thickness set up outdoors should be used for accurate results.

Curtain walls play an important role in building façades of different building types. Glass as a major component, with its reflective appearance, flat surface, and durable properties, present a high-standard building façade to the public. However, due to its transparent properties, glass also has problems in heat transfer. The code of practice for Overall Thermal Transfer Value in buildings 1995 issued by the Buildings Department applies to all hotels and commercial buildings to control the Overall Thermal Transfer Value (OTTV), to reduce heat transfer through the building envelope and the use of air-conditioning. Therefore, different kinds of glass were invented to solve the problem. Since 2015, Residential Thermal Transfer Value (RTTV) applies to residential buildings to control the RTTV of building envelopes, including visible light transmittance and external reflectance of the glazed portions. The RTTV of the wall and the roof should not exceed 14 watts per m^2 and 4 watts per m^2 respectively.

Insulated Glazed Unit (IGU)

IGU, also called double glazing, is widely used in the façade industry. It consists of two glass layers with an inner air space. The distance between the glass surface controls the insulation properties of the glass. It is also a compromise between maximizing insulating value and the ability of the framing system to carry the unit. For more stringent requirements, inert gas, like argon, may replace the air in the space to achieve better performance. On top of the air space, the reflective coating or low-e coating on the glass surface also contributes to the insulating performance.

玻 璃

玻璃是古人類發現最美麗、最珍貴的寶石之一，追朔至銅器時代。今天，玻璃依然是建築設計中重要的物料，可以塑造不同的空間和人造裝飾。

由梳打、石灰粉、矽和其他成份包括鎂和鋁等，都是製造玻璃的主要物質。以原物質燃燒至 1,490 至 1,550℃，可成為半液體的玻璃狀態。玻璃片由不同的方法製造，包括抽壓法、浮動法與滑輪法等。

玻璃的成份須依據 BS952 第一部份：1978 年，以耐用度為主。

— 矽和鋁合共不能少於 71%
— 鹼性（Na_2O; K_2O）成份不超過 15%
— 石灰和鎂合共不能少於 10%

因為厚 6 毫米的玻璃，具有幾乎等於零的視野阻礙和不反光程度，所以經常為香港的住宅樓宇採用。完全透明的玻璃，是以 85% 的透光程度作標準。玻璃幕牆常用的高反光玻璃，如 12 毫米的玻璃，便以透光程度低至 10–20% 作標準。普通的玻璃，因其表面有不同的折射和反光，固較少採用作建築用途。

如要提高玻璃的安全程度，最好是選擇強化玻璃或夾層玻璃。與普通玻璃同一厚度相比，強化玻璃可以提供三至五倍的堅硬程度。而且強化玻璃破碎後，只會產生很碎小的玻璃塊，避免了危險事故的嚴重性。夾層玻璃因為層面間的強大拉力，破碎了的玻璃片不會倒下。而強化玻璃因經過強化過程後，不能隨便切割，故此玻璃的尺寸需預早訂造。在香港，如 Saint Gobain 等製造商，亦開始在本地設廠製造強化玻璃，產品的尺寸更可達 2,600×3,400×19 毫米。

在玻璃技術的設計上，須考慮因熱能改變所造成的裂縫。形成裂縫的原因，是由於暴露於陽光下的玻璃面，受著熱能影響而膨脹，另一面的玻璃面，卻因未達到膨脹的標準而形成玻璃邊產生拉力。當拉力超過了玻璃本身可承受的力度後，便會造成裂縫，尤以大面積玻璃、吸熱玻璃和網線玻璃為甚。預防裂縫的方法包括：

(1) 利用鐵片附在玻璃上可用作隔熱。但以玻璃幕牆為例，玻璃片與片間的距離需要超過 4 毫米以上。
(2) 避免建築物受熱，可以在玻璃片後附加厚簾幕和熱能反射片。
(3) 塗上良好的矽脂彈性膠，可以防止濕氣積聚，以及容許玻璃之間有少量活動。

玻璃承受風壓阻力的程度取決於玻璃的厚度、類別、邊位處理方法以及尺寸大小等。在香港，風壓是以玻璃片距離海港表面水平的尺寸準確計算。如在水平面為 1.2 kPa，但到山頂之時則為 4.3 kPa，若是高層建築物，則需要以模型放在風管作為測試標準。

玻璃的厚度容許含色玻璃有不同顏色的變化。玻璃經表面處理後可有銀、銅、灰、綠、藍、金、粉紅等色。在決定適合顏色時，可以在地盤現場把大塊的玻璃片樣板掛上來觀看其效果。

玻璃幕牆在不同的建築類型中承擔非常重要的角色。玻璃作為主要組成部份，用其可反射的外觀、平滑的表面和耐用的性能成為高標準的建築外牆。然而，因為透明的關係，玻璃的傳熱度高，對於室內的空調要求，成為了一個節能問題。屋宇署 1995 年頒布的「樓宇總熱傳送值」(OTTV) 守則實施於所有的酒店及商業建築上，用來控制總熱傳送值，以減少從建築表面和空調系統帶來的熱傳送。因此不同種類的玻璃及配搭，在市場上正被應用於不同層面，以解決環保問題。自 2015 年開始，住宅樓宇的能源效益設計和建造也需受到監管，目的在於控制「住宅建築物外殼的總熱傳送值」(RTTV)，包括玻璃部份的見光透光率和外部反射率。其設計及建造規定牆壁和住宅屋頂的熱傳送值，分別不可超過每平方米 14 瓦特和每平方米 4 瓦特。

A summary of glass types
玻璃類別的總結

Glass 玻璃	Manufacture 製造過程	Features 特色	Application 用途
1. Float glass 浮玻璃	Made by floating a ribbon of molten metal, up to 38 mm thick 把液體狀的玻璃灌於金屬上，厚度可達38毫米。	Precise flatness, smooth, low distortion 精確的平滑表面，柔滑，微誤差。	General construction, showcases, display windows 一般用途，展覽櫃與廚窗。
2. Sheet glass 平面玻璃片	Produced by drawing glass sheet from a pool of molten glass, traditionally up to 6 mm thick 溶解的液體狀玻璃，由抽壓法製成6毫米的玻璃片。	Transparent glass; the two surfaces are never parallel; some distortion of vision and reflection 透明玻璃，兩塊平面因不能平衡而產生輕微偏差。	General glazing purposes when slightly distorted vision is acceptable 普通用途，並不要求高度平均反光度的玻璃。
3. Rolled and rough-cast glass (translucent) 滑動壓法與不平滑玻璃（半透明）	Pattern engraved on a roller transferred onto the glass surface 把刻有圖案的轉輪壓在玻璃面上成為印花。	Decorative applications, light diffusion and transmission 裝飾用途，較容易讓光通過。	General construction, partitions, bathrooms, doors 用於一般建築，如隔牆、浴室、門。
4. Wired glass (polished or figured) 鐵線玻璃（平滑或不平滑）	Inserting metal wires or a wire mesh into the glass: cross mesh, rhombic mesh, unwired 在玻璃內層，加入鞏固鐵網或鐵條，如交疊成90度的、菱形角度的或鞏固的鐵網。	Fire resistant, prevention of theft and of accidents, due to shattering 具防火功能，亦可防止盜賊潛入或因意外而被打碎。	Internal and external use for construction, shops, doors 室內和室外的施工用途，如商店、門。
5. Heat-absorbing glass 吸熱能玻璃	Glass with colour tints formed by addition of minute amounts of iron, nickel, cobalt, etc. 在混合玻璃液體狀時加入少量的鐵、鋇等，可使玻璃含有輕微色彩。	Controls sunlight heat; colour for design aspects 控制陽光的熱能傳送。同時有不同的顏色選擇，增加保密功能。	External use of construction 室外用的建築用途。
6. Heat-reflective glass 反熱玻璃	Made by plating metal oxide onto surface of float glass; can be made by clear float glass or heat-absorbing glass 在浮玻璃的面層加上氧化鐵物質而產生的。浮玻璃可由透明或熱能玻璃代替。	Interception of sunlight; half-mirror effect, ensures privacy, reducing 融合陽光的光線，具有部份鏡的特質和保密性能，同時亦減低冷凍系統的負荷。	External use in construction, high-rise buildings 用在一般高層的建築物外牆。
7. High-performance heat-reflective glass 高效應反熱玻璃	Monolithically coated with metallic finishes by sputtering method; can also be applied on tempered glass 噴射法可以令附於表面的金屬層平均分布。亦適用於強化玻璃上。	Wide range of colours, mirror effect, energy saving, interior comfort, privacy 有很大的顏色變化和鏡子特性，亦可節約能源，造成舒適的室內空間和保密用途。	External use on construction, high-rise buildings 用在一般高層的建築物外牆。

Glass 玻璃	Manufacture 製造過程	Features 特色	Application 用途
8. Double glazing 雙重玻璃	Dry air sealed between two sheets of glass; the inside surface of the outer glass can be coated with metal film 於兩塊玻璃之間密封空氣，同時在置於外面的玻璃片內附上金屬表面。	Insulation, prevention of dew condensation, pleasant room atmosphere; specify size early 隔熱、防止霧氣形成，讓室內空間更加舒適，但需要盡早決定其尺寸。	Buildings with high heating or cooling needs, or with strictly controlled humidity 用於需要高度熱能或冷凍的建築物，或用於需要嚴格控制濕度的室內空間。亦有隔聲作用。
9. Laminated glass 疊層玻璃	A transparent, tough adhesive polyvinyl butyral film between two or more sheets of glass 以透明、強硬的乙烯聚合物黏劑把兩塊玻璃黏一起。	Safety, prevention of theft 安全，具有防盜作用。	Shopping arcades, doors 商場、門等。
10. Laminated tempered glass 疊層強化玻璃	One or more layer of polyvinyl butyral film laminated between sheets of tempered or alternating sheets of tempered and annealed glass 以乙烯聚合物黏劑把兩塊或以上的強化玻璃黏一起。	Resistance to impact and penetration; for bullet-proof glass 能夠承受撞擊，如防彈玻璃。	Burglar-resistant showcases, glass floors, aquarium tanks, animal cages 防盜用途的展覽玻璃、玻璃地合、水族館的玻璃箱、動物籠
11. Tempered glass 強化玻璃	Made by heating glass up to 700°C and then blowing air onto the surface, cooling it down rapidly and uniformly so a compressed layer is formed on the glass surface 將玻璃燒至700°C，然後以冷空氣快速地使溫度驟降至原來的模樣，玻璃的表面因此變成可受壓的狀態。	Strength, thermal shock resistance, safety 堅硬，可承受熱能的撞擊，是較安全的玻璃。	Doors, balustrade, places requiring safety 門、欄杆、需要安全的地方。
12. Curved glass 弧形玻璃	Soften glass under high temperature and shape it along a mould 把玻璃燒熱軟化在彎形模上而成。	Special aesthetic effects 特別的裝飾用途。	Building façades, corner windows, skylights, display windows 建築物外牆，轉角的玻璃窗、天窗、展覽用途的玻璃窗。
13. 'Colourback' spandrel glass 背有顏色的玻璃	Ceramic coating fused on back of glass to produce a variety of colours and patterns 以陶瓷物質溶解於玻璃底面，可造成不同顏色與圖案。	Many colours and patterns; special aesthetic effects to be custom made 多種顏色和圖案選擇，具有特別設計的外觀裝飾作用。	Spandrel glass, signage graphics 側牆的玻璃片，訊號指引和圖解等。

Saint Gobain Tempering Factory Limited

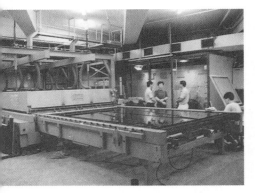
Computerized tempering machine

Heating to 700°C

Blowing air onto heated glass

中空玻璃組件

中空玻璃組件也叫雙層玻璃組件,廣泛用於建築外牆。它由前後兩層玻璃和中間空氣組成。玻璃之間的距離決定了玻璃的隔熱性能。在安裝組件時,隔熱值和窗框承受力要相互妥協。對於更高要求的隔熱,會使用像氫這樣的惰性氣體代替了空氣於玻璃之間,以達到更好的效果。除了靠空氣,玻璃表面上的反射塗層或低輻射塗層,也能幫助提高隔熱性能。

Special installation methods for glass

A suspension glass system is designed to glaze huge openings by suspending large pieces of glass with special metal clamps at the top in a concealed metal frame. This system is applicable to entrance lobbies of office buildings and hotels where a wide view is desired, illuminating the grandeur of the building. Float glass is available from Asahi or Nippon Sheet Glass in heights up to 13 m with thickness up to 38 mm. Delivery and installation on site have to be well planned. Generally, a height of over 6 m requires a suspension system. Glass stiffeners at a right angle to the facial glass are installed and sealed by hi-modulus silicone structural sealant as a bracing support.

Tempered glass and tempered glass doors form a glazing system for wide openings. A specially structured metal glass-suspension unit supports the glass. Either glass ribs or glass stiffeners give lateral support to the glass panels. The glazing height is governed by the maximum tempered glass size produced, which is about 2×4.2 m. A high degree of precision is required for setting up this system.

The largest opening possible to cover by glass is the Pilkington wall or similar structural glazing system formed by 12 mm thick tempered glass panels held together by glass mullions, metal fittings, and silicone structural sealant. Supported by a back structure of elegantly designed steel or concrete frame, very large externally pure glass planes can be constructed.

Glass balustrades are yet another elegant application of glass. Usually 12 mm thick tempered glass is used and bedded a minimum of 100 mm deep into the ground support. Application can be for handrails, parapets, and staircase railings. The structural requirement must meet a minimum horizontal impact load, which is 0.75 to 3.0 KN per m for places of assembly for line lands and depends on the number of people gathered.

特別的玻璃安裝方法

吊懸玻璃的設計是利用暗藏的金屬結構架去承托巨型的玻璃片,通常用於商業大廈和酒店大堂的入門設計,使顯得更堂皇華貴。高度可達 13 米,而厚度為 38 毫米,但在運輸與地盤存放方面須特別小心。一般而言,超過 6 米的玻璃都需要用吊懸法安裝。承托強化玻璃片的結構架以直角的方向接連玻璃面,並要以結構性矽膠輔助接連。

強化玻璃與強化玻璃門形成了大面積的玻璃設計,所以會採用較特別的結構架去承擔玻璃片。一件完整強化玻璃的高度受 2×4.2 米的面積限制。此類結構法要求非常準確的現場安裝技術。

大型的空間部份以「Pilkington 牆」或其他類似結構承托玻璃片,當中 12 毫米厚的強化玻璃片由金屬配件、結構用矽膠及其他玻璃片支

撐。若以鋼架或混凝土架作後層結構，則可造出美觀的外牆全玻璃片格設計。

全玻璃欄杆是另外一項美觀的設計，一般以 12 毫米厚玻璃及結構 100 毫米深的底部支撐，用於扶手、護欄及梯級欄杆等。而用於公眾地方時要承受 0.75–3.0KN/m 的橫向重量，並要視乎行人的流量。

Glass blocks

Glass blocks have many of the attributes of glass, but another dimension is possible due to the thickness of the glass block. Thanks to a variety of sizes, shapes, and patterns, immense visual appeal can be produced. Through translucence or transparency, the designer controls the transmission of light to produce inviting and interesting environments.

Glass blocks are made by fusing two halves of pressed glass together, creating a partial vacuum, which gives an insulating value better than a 300 mm thick concrete wall. Thus energy consumption due to artificial environmental control is lowered. Noise transmission can also be reduced.

Glass blocks can be used externally or internally as a non-structural panel wall supporting no other load from the building. The maximum area of a single panel is limited by regulation to be 9 m^2 and the maximum vertical dimension is 4.5 m for external walls or 6 m for internal walls. The standard size of blocks can be squares of 150 mm, 200 mm, or 300 mm, of 100 mm thick. Waterproofing for glass block walls needs special attention in detailing and workmanship.

The glass blocks are laid on mortar of 1:1/2:4 to 1:1:4 cement, lime, and sand mix, fairly dry. Every third horizontal course is reinforced with expanded metal strips tiled to the adjacent structure. Joints are formed by white cement grout or white silicone sealant.

玻璃磚

玻璃磚具有很多玻璃的性質，而它的「厚度」成為本身的特點。隨著它的形狀、圖案及透光程度，可製造出不同的氣氛與環境。

玻璃磚以兩片玻璃壓製而成，而且內部真空，因此比 300 毫米厚的混凝土牆有更佳的隔熱及隔聲效果。

玻璃磚可用於外內牆。單獨一幅牆的最大面積可達 9 平方米，而外牆最高可砌 4.5 米、內間牆最高可砌 6 米。標準尺寸分為 150 毫米、200 毫米或 300 毫米的方形磚，厚度均為 100 毫米。玻璃磚牆的防水處理需要特別注意細部及施工。

鋪砌玻璃磚用 1：1/2：4 至 1：1：4 的水泥、石灰和沙的砂漿，成份較乾。每三橫列應加鋼片作強化輔助，介面用白英泥漿或矽封料完成。

Curtain wall

A curtain wall as a thin layer of outer skin on a building has had wide application since the early post-war period in the US, when pre-engineered building systems and fast construction schedules were attained. Continuous development has also made curtain walls wholeheartedly embraced by adherents of

Saint Gobain Limited computerized glass cutting

Machine to work glass edge

Machine for openings

Traditional labour cutting for special shapes

Large plane of glazing at the Hong Kong Convention and Exhibition Centre

Frameless glazing in the lobby of Greenwood Garden, Sha Tin

Glass roof for walkway

Glazing in the lobby of the Exchange Square

Glass enclosure for the staircase at the HSBC headquarters

Glazing details at the Hong Kong Convention and Exhibition Centre

Pilkington glass system at the Hong Kong Club

Glass roof at the Hong Kong Park

The Bond Centre with stick system

The City Hall, Central, with an early form of glazing system

The colourback glass of the Sun Plaza Complex in Greenwood Garden, Sha Tin

Granite cladding of the Exchange Square

The Ruttonjee Centre

The monolithic, reflective glass of the Vicwood Plaza

Pacific Place

The New East Ocean Centre

Glass block wall

Curtain wall installation (stick system)

The Bond Centre

postmodernism, as a flexible palette for postmodern expression. The wave of curtain walls also struck Hong Kong in the 1980s, when high-rise matchbox architecture of curtain walls evolved into the present sculptural and colourful forms expressing state-of-the-art technology. The early conventional curtain wall is monotonous and often criticized as lacking human scale. However, with the introduction of new technology such as metal cladding, stone finishes, back colour coating to glass, and glass fibre panels, the aesthetic and technical virtues of curtain walls are extolled.

Curtain walls can be classified according to their system of installation:

1. The stick system (e.g., the Lippo Centre) is installed by first erecting vertical mullions, followed by horizontal transom members. Glass or metal forming the spandrel are installed next. Then the window units or vision glass panels are fixed into place. This system was widely used in the early stages of the history of the curtain wall and is still used by some manufacturers. It also has the benefit of economy due to minimal bulk in shipping and handling costs. Some dimensional adjustments to site conditions are possible. Its limitation is that the system has to be assembled on site and has difficulty taking on stone cladding. As the installation is done piece by piece, the installation process cannot be too fast.
2. The unitized system (e.g., the Exchange Square) is made up of large framed units pre-assembled at the factory with spandrel panels and glazing. The perimeter edges of the units join and interlock in a more horizontal mode to produce a continuous building enclosure. As much of the pre-assembly work is done at the factory, site fixing can be very fast, which is an advantage in tall buildings. Also, stone finishes are viable.
3. The column cover and spandrel system is formed by column cover sections together with long spandrel elements spanning between structural columns. Infill glazing units can be pre-assembled in the factory or fixed on site. This method is a true expression of the structural frame of a building.

Structurally, the curtain wall is subject to the action of wind, which produces positive and negative (suction) forces on the building. Unless there is a previous test history, the Building Authority requires either the cyclic (positive and negative pressure) test or the static load test to be carried out on a specimen of at least one floor high and which includes the features of the curtain wall being used. During the test, the glass should not break, nor should the deflection of any structural elements of the specimen exceed 1/180 of the span or 20 mm, whichever is less. Fifteen minutes after the test load, the extent of recovery should be at least 95%, and there should be no sign of plastic deformation, significant separation, or deleterious effect.

The requirement of government submissions for curtain walls includes structural details with erection procedures and calculations to demonstrate adequate strength and stability. A typical section with structural members and supports together with connection details has to be submitted. Electrolytic action should be prevented.

Technically, the movement of curtain wall cladding elements relative to one another has to be considered. These may be the results of thermal expansion or torsional movements of the floors. As the external skin is under the direct influence of external climatic conditions and the internal layers are affected by the interior microclimate, local variations in temperature will affect the curtain wall. It is thus desirable to maintain the glass in a somewhat freely supported position at a certain degree of tolerance so that no adverse reaction can result due to movements in the frames or the building structure. The anchor attachment of the curtain wall to the support structure may be designed to allow three-dimensional movement, providing a flexible restraint. It is noted that structural stability, movement, and watertightness are essential design considerations for curtain walls.

Another aspect is protection against fire spreading from storey to storey. A fire-resistant material, usually fibreglass with a minimum one-hour FRP, is used to fill the cavity between the edges of concrete slabs and spandrel glass or cladding. The exact FRP follows part of the concrete slab.

A practical consideration is the need for proper protection of glass surfaces during erection, as materials and contaminated water like concrete wash-off, weld splatter, and wind-blown debris may cause surface damage and discolouration of glass coating.

Finally, a gondola system for maintenance and cleaning of the curtain wall system has to be provided and pre-planned for installation. Gondola tracks and storage space for the gondola need to be provided at the design stage to cater for the loading and the space thus occupied.

The Exchange Square

玻璃幕牆

八十年代的玻璃幕牆浪潮亦影響到香港建築由早期的平板設計走向現代科技，塑造出各種特別的形態和顏色。玻璃幕牆向來被批評為缺乏「人味」的設計，但混合了金屬板模、石塊片、噴色玻璃、玻璃纖維板等技術，幕牆式設計再不是千篇一律、平淡無奇了。

幕牆類別是根據安裝方法大致分為下列幾種：

(1) 柱架組合法（如力寶中心）是先安置垂直的結構架物料，再嵌接橫向的部份；隨後再裝上以玻璃或金屬覆蓋板的不見光部份板塊，最後把透光部份的玻璃裝上。此類別的玻璃安裝法是初期幕牆最常採用的，由於物料本身運送輕便，因此也是較為便宜的方法；而且原物料在地盤安裝時容易切割，能配合實際情況上的偏差。但這種安裝方法卻不適合於石塊覆蓋板的裝嵌，原因是所花的施工時間較長。

(2) 單元組合法（如交易廣場）是把整體的架構單元分為不通光部份與通光部份，並先在工廠內加工與裝嵌。單元部份留有鉗接的組合關件，再於地盤安裝。因此，在地盤上的安裝工程也較為簡單和快捷，特別是用於高層建築物時，則更為理想。此類方法更可裝配石塊的板模。

The Leisure and Cultural Building in Ping Shan, Tin Shui Wai

(3) 結構柱的表面板塊裝嵌與不通光板模的組合法，先以長型的組合料件安置在結構柱或陣樑上，組合的結構架間嵌置玻璃片塊，可在工場裏預先製造或在實地切割而成。這類方法為建築物的結構架作真實表現。

玻璃幕牆的結構，須能承受建築物的風壓力與相應產生的吸力。除非備有過去同類測試報告的證明，否則必須呈交建築事務監督有關的周期性測試（正與負的壓力）或靜止的負重測試，而所需的測試報告必須包括在一層樓高度的幕牆。在測試中，玻璃片不能裂開或結構料的偏差不能超過 1/180 的結構跨度或 20 毫米（以最少者為準）。測試後 15 分鐘，整件物料架應有超過 95% 維持原狀。

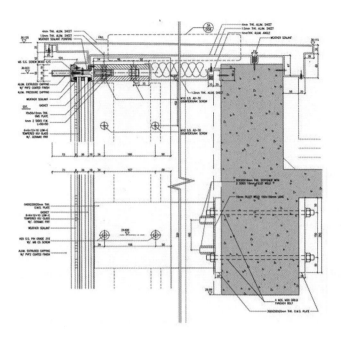

The HSBC headquarters: metal cladding with flurocarbon coating

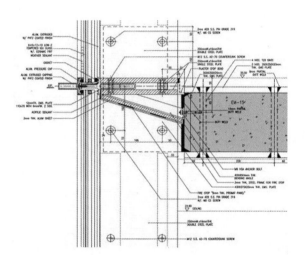

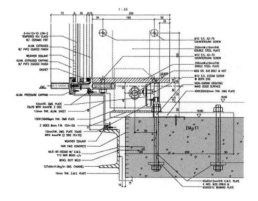

Glass box details

玻璃幕牆的設計須經建築事務監督批核，才可以於地盤施工。呈交的報告包括結構的細部設計圖、施工程序、結構設計的計算細則等，以證明設計的幕牆合乎安全標準。設計須避免產生電解作用。

技術上，應注意玻璃幕牆設計的熱能效應所產生的相對性活動，或樓板扭力的活動。周邊建築物的表面，因外間溫度與室內溫度有偏差，造成了微氣候的變化而影響了幕牆。因此設計玻璃幕牆時，必須連同前述的影響因素一併考慮。至於建築物周邊的結構，必須要計算其立體活動的空間。除了結構方面，移動與防水等因素，同樣是設計玻璃幕牆時應要留意的問題。

其次，必須防止火焰從肇事樓層傳送到別的樓層去。一般設計以具有最少一小時防火物料，填塞陣的空間，使混凝土面與外牆的板模間，再沒有可傳遞熱與火的空間。實際的防火時間乃根據混凝土樓板防火需要。

施工時，須注意保護玻璃表面，並清洗積聚於玻璃面上的污水。例如起板、灌注混凝土，又或打風吹起的垃圾及污水，均會破壞玻璃的表面；應加強保護，防止玻璃表面侵蝕或褪色。

最後，還要考慮清潔玻璃幕牆所用的吊船系統裝置。在整個設計項目中必須預留空間作收藏機械之用，以及考慮吊船移動的路軌系統設計及承重量。

Skylight at the entrance of the Hong Kong Club

Skylights

Psychologically, daylight is believed to be more pleasing than is artificial illumination. Skylights are used to introduce this visual and psychological connection with nature. Development with this daylight treatment seems to be more popular nowadays. Naturally lit atriums and roof-glazed malls have become internal streets for people.

Skylights provide daylight transmission and weather protection to roof structures. Skylight modular units are usually prefabricated at factories and come in a variety of shapes such as dome, pyramid, trapezoidal, and barrel vault. Materials range from acrylic, unreinforced PVC, wire-reinforced PVC, glass fibre to high-impact-resistant polycarbonate with colours of clear, tinted, and opal. Installation comes with non-combustible metal curbs.

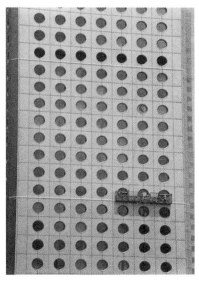

Aluminium cladding of the Jardine House

A combination of glazing with a metal structural frame can produce skylights with a large span. Structures like space frame, trusses, vaults, etc., are developed alongside the large space. Material for glazing can be laminated glass, tempered glass, acrylic, or polycarbonate. Laminated glass is safe for use. Tempered glass is strong. Acrylic or polycarbonate is widely used in barrel-vault-type structures. Double glazing is used for better safety and insulation. Shading devices can also be incorporated in the skylight design.

In designing the skylight, various factors are considered: function for lighting or aesthetics, degree of natural light admitted, effect on other building services like air-conditioning and artificial lighting, type of glazing, span and details of the structural support, sealants and gaskets, detailing of joints between supporting structures, any expansion joints, future maintenance, and site-erection procedures, and delivery.

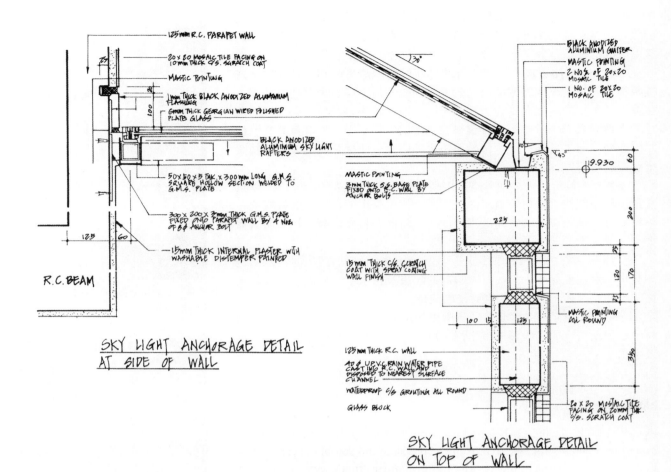

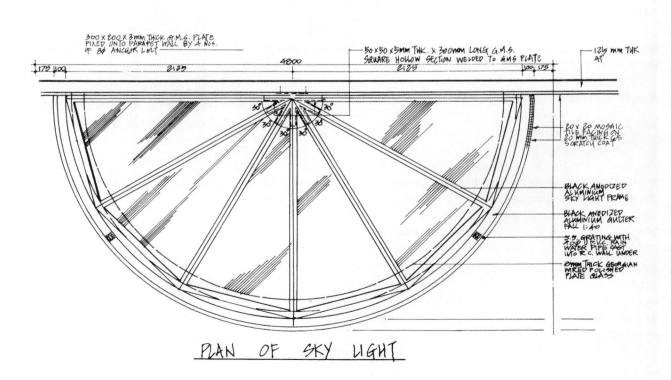

Details of a skylight

天 窗

心理上,天然光線遠比人造光線更為自然和舒服。天窗的特別設計靈感,也是讓人恍如置身於自然的感覺,這種設計漸漸地普及,為許多中庭及以玻璃天窗做屋頂的購物商場,塑造出自然「街道」的感覺。

天窗既可傳送日光,又可保護室內的環境。天窗的模式組合結構,通常在工廠裏加工和預備,包括了特別的形狀如圓頂、金字塔、錐形、梯形、連續半圓拱等。物料類別包括塑膠、鋼線結構塑膠、玻璃纖維以及高阻力纖維膠等,還有不同的顏色度如透明的、調色的、乳色的。裝嵌時要以非燃性金屬框作模架。

混合透光玻璃與金屬結構架可提供較大的天窗設計。結構系統中的立體構架、樑架、拱架等常用於大型的空間設計,當中的透光物料包括多層重疊玻璃、強化玻璃、塑膠或纖維化聚碳酸脂。多層重疊玻璃較為安全,而強化玻璃是比較堅固的物質;塑膠或聚碳酸脂較多應用於連續的半圓拱結構。雙重玻璃片可加強安全、防熱或隔聲等功能。在天窗的設計中,時常會加入活動遮光的裝置,使空間添加彈性和舒適感。

設計天窗時需考慮下列幾項要素:

(1) 照明和美觀的功能;
(2) 對天然光程度的要求;
(3) 其他建築設施,如空調、人工照明系統、透光片物料的性質;
(4) 結構的跨度和設計性質;
(5) 封料和墊料;
(6) 結構架之間的細部設計;
(7) 伸縮介面的設計;
(8) 維修設備;
(9) 在地盤實地安裝和運輸。

Installation of a 20 m diameter skylight at the Laguna City

2.12
Floors, Walls, and Ceiling Finishes
地台、牆身、天花物料

General

Selection of the type of finish depends on a number of factors such as the function of the area, the type of base, degree of comfort required, appearance and consistency with the overall design, budgetary limit, maintenance, safety, and individual preference of the architect or owner. Types of finish can be in situ or prefabricated, rigid or flexible, and of a variety of building materials available in the Hong Kong building market.

概 述

物料的選擇由許多因素決定，包括空間功能、底層部份的處理方法、舒適程度的要求、外觀、整體的設計意念、經濟預算、維修保養、安全程度、建築師或業主本身的個人喜好和選擇。物料的類別則取決於市場的供應、可否在地盤加工，或在工廠完成後直接運送到現場。

In situ floor finishes

In situ floor finishes are mixed on site, laid in a liquid or semi-liquid state, then dried and set as a hard jointless plane.

Mastic asphalt is impervious to water and forms a good surface to receive thin tile and flexible sheet finishes, applicable to utilities areas and washrooms. The application of asphalt is discussed in section 2.7 on roofing, waterproofing, and expansion joints.

Granolithic finish as a mix of cement and granite chippings comes in several finishes:

1. Trowelled: smooth or wood float finish
2. Rubbed: rubbed down with carborundum stone to expose the chippings
3. Washed: before hardening, the surface is brushed off to expose the aggregate.

For trowelled or rubbed granolithic, a 2:1:4 cement, fine and coarse aggregate mix is used, the colour depending on the aggregate or stone chippings selected. For a washed finish, a 1:2 cement and aggregate mix is used. The thickness for laying on hardened concrete is 40 mm. To obtain a coloured finish, coloured cement is used as a 10 mm thick finishing coat and applied before the undercoat has set.

Terrazzo incorporating marble aggregate can produce a variety of colour finishes such as for lobbies and corridors. Laying is in two coats. The undercoat 10 mm thick is of 1:3 cement and sand mix. The finishing coat 15 mm thick is of 1:2 to 1:3 coloured cement and marble aggregate mix, the bigger the size of the aggregate, the more the proportion of which is required. Terrazzo work is carried out in panels of 1 m². The surface is grinded to expose the aggregate, forming a smooth finish.

Floor screeds with various textures can be used as floor finishes for low-cost housing, factories, warehouses, car parks, etc. This is discussed in section 2.13 on plasterwork.

Dividing strips of brass, stainless steel, or plastic 3 mm thick and the depth of the screed is used to separate different floor finishes like terrazzo and granolithic.

When a smooth floor finish is extended to form ramps and steps, non-slip strips of 1:1 cement and carborundum dust mix are provided, projecting with a curved top about 4 mm from the finished floor.

Washed granolithic finish

Rubbed granolithic slabs

Terrazzo slab

在地盤組合的地台物料

地盤的實際施工中，在原地組成的物料要先混合於液體或半液體的狀態中，然後風乾，待凝結後便組成了堅硬的地磚平台。

非滲透性的地瀝青膠漿是安放薄石磚或彈性地台的最佳物料，常應用於工作間或洗手間等地方。瀝青的應用和特性，本章第 2.7 節〈屋頂的鋪工、防水層、伸縮縫〉中有詳細介紹。

人造石物質主要由水泥與麻石的碎粒組成，碎粒是經：

(1) 鏝光面：平滑或用木條推平。
(2) 磨面：用金鋼砂石磨平，把細石粒表露出來。
(3) 水磨：在人造石未凝固前，先洗去表面的水泥，使石粒外露出來。

鏝光面或磨面人造麻石中的水泥、微粒和粗粒的比例為 2：1：4，石的顏色取決於粒料或石粒子的成份。至於水磨石的水泥與粒料的比例為 1：2。鋪砌於堅硬混凝土的厚度約為 40 毫米。而為了製造特別的顏色，可以用 10 毫米厚的顏色水泥作表面處理方法，但要在底層未凝固前裝置。

水磨石包含了雲石的粒料，並有許多不同的顏色，分別應用於大堂、走廊等地方。鋪砌時以兩層為主：

先以 10 毫米厚 1：3 的水泥、沙石作底層，完成後再以 15 毫米厚面層 1：2 至 1：3 的顏色水泥沙與雲石沙粒混合。如雲石較粗，則雲石沙粒比例要加。水磨石工程是以面積 1 平方米為板塊模數，表面經打磨後，再露出平滑的粒料。

水泥砂漿可以有不同的質感，迎合不同地台的需要，如公共房屋、工廠大廈、貨倉、停車場等。這在第 2.13 節〈抹灰工程〉會有討論。

可用 3 毫米厚的鋼片、不銹鋼片或塑膠片，分隔不同物料的地台，如水磨石等。

Stockpile of ceramic tiles

Artificial granite tiles showing keyed base

Ceramic tile fixing

Applied floor finishes

Applied floor finishes are supplied in tile or sheet form and are laid onto a suitable base, usually of screed.

Ceramic and quarry tiles are a popular class of floor finishes and have a vast variety of patterns, colours, and sizes. Surfaces range from unglazed, semi-glazed, glazed, to highly polished finish. Tiles are classified by the method of manufacturing, such as extruded tiles, dust-pressed tiles, and cast tiles, and the water absorption limit. Properties of tiles are specified by scratch hardness, water absorption, crazing resistance, acid resistance, thermal shock, etc. Applications of ceramic tiles include in lobbies, bathrooms, kitchens, and balconies.

The range of deviation in the facial sizes of tiles in individual consignments does not exceed 0.5%. Deviations in the work size thickness do not exceed ± 0.5 mm.

The method of testing by sampling tiles regarding warpage, curvature, water absorption, resistance to crazing, resistance to chemicals, and resistance to impact can be obtained in BS 1281:1974. The warpage of tiles is not to exceed 0.5 mm. The curvature of tiles is not to exceed a concavity of 0.13 mm or a convexity of 0.76 mm. The index of wedging is not to exceed 0.08 mm per 25 mm. Water absorption is limited to 18%.

地台磚施工

鋪砌地台的物料，多為磚塊或薄片，施工時以砂漿料鋪好底層。

磁磚和黏土瓦片是常見的地台物料，有許多不同的形狀、顏色和尺寸。表面由非光面、半光面、光面到高度的滑面。磚瓦分類以其施工程序和吸水程度來辨別，如擠壓磚、塵壓磚、鑄瓦等。磚片分類視乎受磨硬度、吸水程度、受擠壓阻力、對酸性和熱能的反應。磁磚多用於大廳、浴室、廚房和露台等。

至於磚片形狀的大小與偏差則不能超過 0.5%，且物料厚度亦不能超於 ±0.5 毫米反差。

磚瓦樣板測試包括翹曲、彎度、吸水程度、受壓程度、化學反應、撞擊反應等，可參考英國標準報告項目 BS1281：1974。磚瓦翹曲不能超過 0.5 毫米，受彎程度不能超過凹面的 0.13 毫米或凸面的 0.76 毫米，斜角度標準則定為每 25 毫米為 0.08 毫米以下，吸水能力亦只可以在 18% 之內。

Fixing of ceramic floor tiles can be done by the following methods:

1. Semi-dry method. The concrete base is adequately true and level, to achieve specified tolerances and be free from contamination. Tiles are soaked in clean water to allow for absorption and expansion. Semi-dry 1:4 cement and sand mix is thoroughly compacted to the required thickness, 20 mm minimum, and laid to fall. Cement and sand slurry is poured over the bedding and trowelled to 3 mm thick. Tiles are tamped firmly into the bed, forming an even surface. Joints are set to be regular, truly aligned, and grouted with 1:1 cement sand mix. Surplus grout is cleaned off as work proceeds.
2. Thick-bed method. Screed is laid on the concrete base and allowed to set. The tiles are soaked in clean water, and the screed is damped to

reduce suction. A 15 mm thick 1:3 cement and sand mix is laid. The back of each tile is coated with slurry before fixing. The tiles are then tamped firmly into the bed with well-set-out joints, which are then grouted.
3. Thin-bed method. A bed of adhesive can be used for fixing in accordance with the manufacturer's recommendations. Proprietary grout for the joints can be used. This method is used in particular conditions such as a metal or a timber base or area, allowing for movement.
4. Quick-dry floor screed. The demand for faster hardening of floor screeding rises as labour and time costs have gradually increased in recent years. Pre-packed floor screed products that accelerate the hardening and drying process were invented and have been widely used in the market in recent years. Usually the hardening time is around 3 to 21 days but allows light traffic in 3 to 4 hours. It is preferable in some renovation work like shopping malls. The workers can lay out the screeding at midnight, finish tiling in early morning, and immediately reopen to the public.

Timber block flooring

Roof tiles for flat roofs can be precast concrete tiles of 1:2:4 mix or canton tiles, size 30 or 35 mm thick and 300 or 400 mm², Laying is by bedding on 1:4 cement mortar. Joints are filled with hot bitumen.

Flexible PVC flooring is produced by a mixture of polyvinyl chloride resin, pigments, and mineral fillers as 300 × 300 mm square tiles or in sheet form up to 2,400 mm wide. Thickness is 1.5 to 3 mm.

Laying is done on a flat and smooth screed, cleaned and free from contaminants. Adhesive is applied in a thin film and spread evenly according to the manufacturer's instructions.

Carpets are made from nylon, acrylics, and wool in a vast range of styles, patterns, and colours. To be long lasting, carpets are laid over an underlay of felt or latex and secured by adhesives and nailing. Carpet tiles are available and laid without the use of adhesives, relying on edge fibres to interlock.

Granite finish at the Standard Chartered Bank

As a deluxe form of floor finishing, marble or granite paving slabs of 20 mm thick are laid on solid screed with bedding of 12 mm thick minimum 1:3 cement sand screed. This wet fixing is appropriate for floor slabs and low-lying planters. A white cement base is used with a light-coloured marble. Marble and granite is discussed in section 2.6 on masonry and granite/marble works.

鋪砌地台瓷磚方法可分為：

(1) 半乾鋪砌法
混凝土底層平水須小心燙至可容許的偏差範圍內。同時，混凝土面須保持潔淨。先將瓦片浸於清水中，使吸水和膨脹。半乾的水泥比例為 1：4，加上沙底再混合和擠壓至 20 毫米的厚度，鋪砌瓦片時須維持斜度疏水。澆灌水泥與砂漿於底層之上，再抹平至 3 毫米的厚度，接著把瓷磚穩固地鋪排於其上。全部接駁的部份要有系統地組成一個平面圖案，清潔剩餘地沙泥漿後，可以清楚看到地面的圖案。

Raised flooring at the Printing House

(2) 厚底層鋪砌法
先用沙泥鋪於混凝土面，待其乾涸。把磚瓦浸於清水中，砂漿再次弄濕，有助減低吸水的特性。以 15 毫米厚 1：3（水泥比沙）砂漿鋪好地台，把磚瓦鋪砌其上，待凝固後洗清膠漿。

(3) 薄底層鋪砌法

根據供應商的商標指示，用一層粘合劑作為磚瓦的黏劑，亦可用於連接口。這種鋪砌法多用於特別情況，如底層為金屬或木，又或某些需要擴張、收縮的地方。

(4) 快乾地板砂漿層鋪砌法

隨著近年來香港建築成本及人工逐漸增加，對於地板砂漿底層要求乾透時間更快速。以往較貴價的預先包裝地板砂漿產品也開始被廣泛應用，原因是這種產品加速了底層硬化和乾化的過程。底層硬化的時間通常在 3 至 21 天之間，但 3–4 小時後就允許輕度人流。由於時間對於很多翻新工程來說比較關鍵，所以很多商場翻新時都會使用這種產品。工人會在夜晚鋪上砂漿層，到第二天早上就可以完成鋪磚並即時向公眾開放。

屋頂用的平面磚，可以是特別製造的 1：2：4 為混合系數比混凝土磚瓦，或「Canton」磚瓦，尺寸由 30 至 35 毫米厚與 300 或 400 毫米方。鋪砌時可用 1：4 水泥砂漿的底層，連接口用熱瀝青液。

伸縮性塑膠地板由聚合氯化脂、色料與礦物料等混合製成，為 300×300 毫米的正方形瓦片或至 2,400 毫米闊的大塊片，厚度由 1.5 至 3 毫米。

鋪砌須在平滑的砂漿底層上，而且要保持清潔與不能沾塵。使用黏劑時要依據來源供應商的說明，保持薄與均勻。

地氈可由尼龍、塑膠纖維或羊毛等織成，分別有不同的形式、圖案和顏色。為著保持持久性，地氈可鋪於油毛氈、橡膠乳氈之上，再用黏劑或釘牢固。同時亦有地氈片塊，鋪砌時不用任何黏劑，單靠邊旁的連接位便行。

較豪華的地台物料，有雲石、麻石片，一般為 20 毫米厚，鋪砌用 12 毫米厚 1：3 水泥砂漿底。這種濕鋪砌方法應用於地台或花槽。白色英泥底用於較淺色的雲石工程。另雲石、麻石的應用，可翻閱 2.6 節〈石工、雲石和麻石〉。

Timber floor finishes

Timber floor finishes with a natural wood finish give a warm, comfortable feeling. A resilient floor can also be achieved. Boards, sheets, and blocks of various timber types are laid or attached on a suitable subframe or base. Application is for residential flats, ballrooms, sports halls, squash courts, shops, etc. If the wood block floor is on a ground slab, a continuous damp-proof layer has to be provided against rising damp and preventing the wood to contact ground concrete. Timber floors are discussed in section 2.8 on carpentry, joinery, and ironmongery.

木地板

木地板給予人一種溫暖、和諧的舒適感，且富有彈性。板條、板槐或整片的木，來自不同類別的木質。木地板一般用於私人住宅、球場、運動場、壁球場、商店等。

如需在地面上加木地板，必須先鋪上完整的防水層作保護，防止潮濕空氣由地下滲出，更可隔開混凝土與地板。第 2.8 節〈木工、細木工、五金〉中有關於木地板的詳細討論。

Raised flooring

Raised flooring is a new system for flooring based on a flexible space concept installed for office interiors. Most mechanical and electrical services can be installed, including cable trunking, communication networks, and air-conditioning systems. The main advantage is the ease and speed of changing the service location on refurbishment of the office. For installation of air-conditioning, a height of 250 mm is sufficient. If air-conditioning is not required, a height of about 75 mm is suitable for the installation of a raised floor.

架空地板

架空地板的設計，提供了辦公室中彈性活動空間的新設計意念。當中的有關建築物機電設施都是收藏在管道內，如電訊系統、冷氣系統等。它的優點是更有效率、快捷去更換設備的位置，改變辦公室的設計。在安置冷氣系統時，只需 250 毫米的高度便已足夠；另在不需要安裝冷氣設備的情況下，則只需 75 毫米。

Wall finishes: Wall tiles

Glazed ceramic tiles are common internal wall finishes for bathrooms, kitchens, lobbies, toilets, etc. The base is of wall screeds 10 mm 1:3 cement sand mix with the surface lightly scratched to form key.

Methods of fixing wall tiles to wall screed are as follows:

1. Thick-bed method. Tiles are soaked in clean water and allowed to drain off. The wall screed is damped to prevent absorbing water from the bedding mortar. The back of each tile is buttered with 10 mm 1:3 cement sand mortar. The tiles are tapped firmly into place, and any adjustment is made within ten minutes of fixing. Joints are grouted with 1:3 cement lime mix, surplus grout cleaned off.
2. Thin-bed method. A bed of adhesive can be used for fixing, in accordance with the manufacturer's recommendations. Grouting can be by use of a proprietary grout.

Mosaic tiles for wall finishes can be glass mosaic tiles (vitrified) and glazed ceramic tiles for external wall finishes. Unglazed vitreous mosaic tiles are commonly used for walls and floors of fire staircases and service rooms. Fixing of mosaic tiles is by thick-bed method direct to concrete base, thick-bed method to wall screed, and thin-bed method, all similar to the methods used in fixing glazed floor tiles. However, mosaic tiles come with a backing paper that is removed after fixing. Grouting is by cement and powdered limestone 1:3 mix. Adhesive allowing for flexibility in movement is a good fixing method, particularly for high-rise buildings.

Movement joints provided on the external wall are discussed in section 2.7 on roofing, waterproofing, and expansion joints.

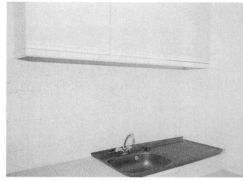

Ceramic wall tiles at the Laguna City, a residential development

Cladding wall and ceiling and granite floor at the Bond Centre

The Hong Kong Cultural Centre

Marble wall at the Windsor House, same pattern for the ceiling and the floor

Metal soffit at the Bond Centre

Good workmanship tiling at the Hong Kong Cultural Centre

Ceramic wall tiles at the New Town Plaza, Sha Tin

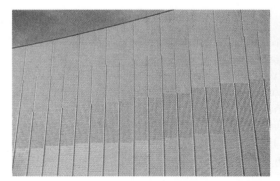

Wall tiles at the Hong Kong Cultural Centre

牆身處理：瓦片

在牆身鋪砌瓦片的程序如下：

(1) 厚底方法

將瓦片侵於清水中，再讓水流乾。然後弄濕牆身的水泥底，避免吸去打底水泥漿的水份。瓦片背面，先塗上 10 毫米厚 1：3 水泥漿，然後把它牢固於適當位置，這時仍容許有 10 分鐘時間再移動瓦片的位置。瓦片間介面可用 1：3 水泥漿封好。

(2) 薄底方法

薄底層是用供應商指定的黏貼漿把瓦片黏在牆身上。

馬賽克可分為一般牆身用的玻璃瓷磚，及用於外牆的光面瓷磚。另外，啞面玻璃瓷磚則常用於走火通道樓梯間或設備房間。鋪砌馬賽克牆身時，需用厚底方法。由於馬賽克出廠時是以一塊底紙固定，所以鋪砌後，須把底紙除掉。再用 1：3 水泥粉末為介面灌漿。特別在處理高層建築物時，要留有可容許建築物移動的伸縮縫。

有關外牆伸縮縫的設計已在第 2.7 節〈屋頂的鋪工、防水層、伸縮縫〉中討論。

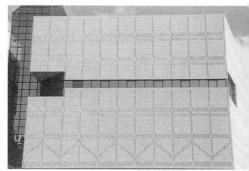

Artificial granite flooring

Wall finishes: Marble and granite slabs

For internal wall finishing, marble or granite slabs 20 mm thick are fixed not less than 12 mm from the structural surfaces on Keene's cement or plaster of Paris dabs. Each slab is fitted with copper cramps and 'S' hooks to the structure in addition to the dabs.

Installation of external wall granite and marble is discussed in section 2.6 on masonry and granite/marble works.

牆身處理：雲石與麻石層

在室內的牆身設計中，若以 20 毫米的雲石或麻石層片鋪砌牆身，須以不少於 12 毫米的堅力士水泥或馬黎灰泥塗於結構層上。每塊石片須用銅夾及「S」型排鈎藏於結構中。

外牆用麻石與雲石的安裝方法已在前文中談論過。

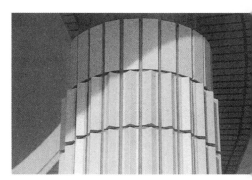

Special tile for circular column

Wall finishes: Dry lining techniques

Walls, especially internal ones, can be finished by dry lining techniques, reducing the water consumption amount for construction. The techniques can be performed with a variety of materials, which can be self-finished or ready for a final finishing plaster. Materials used include plywood, hardboard, chipboard, and plasterboard. Dry linings are fixed by nailing onto timber grid battens, a standard one being 300 mm centres. Plasterboard can also be fixed to a solid background by use of dots of bitumen-impregnated fibreboard secured to the background by board finish plaster.

Screed flooring

Granite interior at the Laguna City

Spray paint at the Hong Kong Club

Timber floor at the Dynasty Club, the Hong Kong Convention and Exibition Centre

Tiling work at the Laguna City

牆身處理：乾襯砌法

室內的牆可以用乾襯砌法，以減低施工時的用水量及表現物料本身的自然物質，如夾板、硬板槐、石屑板和石膏板。

乾襯砌法是利用鐵釘釘固在木塊方格架上，以 300 毫米為標準的方格距離。

抹灰板可裝在堅硬的底板，可塗上瀝青漿的纖維板及灰漿作穩固背面。

Wall finishes: Miscellaneous

Rendering, plastering, and painting are discussed in section 2.13 on plasterwork and section 2.14 on painting.

Fair-faced concrete is discussed in section 2.3 on concrete work.

Wallpaper, because it is easily damaged by other trades, is not incorporated in the course of construction of the building but more often included in interior decoration work.

牆身處理：其他類別

粉飾批盪、抹灰與油漆等會在第 2.13 節〈抹灰工程〉中詳盡介紹；而清水混凝土的牆面，可參閱第 2.3 節〈混凝土工作〉。

裝飾用牆紙較容易受損，只適用於室內設計。

Suspended ceiling

A suspended or false ceiling forms an integral part of the present office interior, as all building services, like lighting, air-conditioning, and fire protection, can be housed inside to give a neat plane ceiling. The membrane of a ceiling system may be required to produce a predetermined amount of sound absorption, thermal insulation, sound reduction, fire protection, resistance to diffusion of water vapour, or a decorative finish.

A large variety of products constructed from different materials are available. Mineral fibre ceilings are common and have an acoustical advantage. Modern technology has improved this type of ceiling to stand very humid external conditions. Aluminium panels are light but rigid, easy to install, and come as a modular design to incorporate lighting diffusers, air intake and outlets, access panel for servicing, etc. Electrostatic powder coating gives almost any colour.

A suspension structure is hung by hot dip galvanized mild steel (GMS) angles from the ceiling to support the GMS suspension channels. Aluminium runners and tee-bars carried by these channels can support the ceiling panels. The appearance of the system can be concealed, exposed, or semi-concealed. Common modular grids are 600 × 600 mm or 600 × 1,200 mm.

Special consideration for particular conditions as listed in CP 290 include: pest infestation, bacterial contamination, chemical pollution, dust-laden atmosphere, dust-free atmosphere, high humidity, vibration, extra protection against impact in occupation, special loading requirements such as wind or access, and expansion joints.

懸吊天花板

懸吊天花板可以遮蔽天花上的建築機電設施，包括燈光、冷氣系統、消防設備等，是現代辦公室常見的設計。

天花板融合了吸音、抗熱、隔音、防火、防潮或裝飾設計等功能。

俗稱為「假天花」的天花板模，可以由不同的物質製造。礦纖維天花較常應用於隔音等功能上。在現代技術改良下，天花板更可應用在高度潮濕的戶外環境。鋁質天花板較輕盈與堅固，安裝亦比較方便，更附設配件如燈飾天花板槽、空氣調節網架、活動板塊片作維修保養門縫等，還可電鍍不同的顏色。

懸吊的結構由浸鉛水生鐵角條組合，從鋁質的橫向槽架伸展 T 型的直向掛伴把天花板牢固空中。懸吊架構分為蔽式、展露式或半展露式。較普遍的組合模組為 600×600 毫米或 600×1200 毫米。

特別的設計用於獨特的環境要求，如 CP290 中列出的包括：防霉、防蟲、化學污染、防塵、高度潮濕、避震，以及特別的負荷如風力、可移動格、伸縮縫等。

Ceiling finishes: Miscellaneous

Dry lining techniques discussed for use in wall finishes is applicable to ceilings.

A plasterboard base secured by timber battens to the structural ceiling can be finished with a skin coat of plaster. Usually timber moulding is added at the junction between ceiling and wall, to avoid any cracking.

Internal plaster and painting for ceilings is discussed in section 2.13 on plasterwork and section 2.14 on painting.

天花的處理：其他類別

在牆身一節中討論過的乾襯砌方法，亦可應用於天花裝置。石膏底板用木條釘在結構天花板底，再塗上薄灰作表面處理。同時，木製裝飾天花線可遮蔽天花與牆身交接位，亦可避免罅隙。

室內的抹灰與油漆工作可參閱第 2.13 節〈抹灰工作〉和第 2.14 節〈油漆工程〉。

Epoxy spray painting

Suspended metal grid ceiling

Metal mesh ceiling at the Science Museum

2.13 Plasterwork
抹灰工程

Application and background

Plastering is the intermediate coating of building materials applied on the internal façade of concrete work or blockwork to create a smooth, flat surface ready to receive paint, wallpaper, or other internal decorative finishes. Rendering is the intermediate coating for the external wall to prepare the surface for external finishes such as glazed tile, mosaic tile, or stone-like spray paint.

The background is prepared by removing efflorescence, laitance, oil, grease, all traces of release agents, dirt, and loose material. At junctions of different backgrounds, such as cast concrete and blockwork, a strip of expanded steel lathing extending 150 mm in width on each side should be nailed down before the application of plaster. For vertical concrete surfaces and soffits, splatter dash is applied immediately after the striking of formwork (within four hours). This is a 1:2 to 1:2.5 mix of cement and coarse sand mixed with water to give a consistent mortar, which is then dashed on with a roughcast uneven appearance and is kept wet for about two hours. After wiring and hardening, the undercoat of plastering or rendering can be applied.

For hardened or existing concrete walls, the base is prepared by hacking off concrete to a depth of 3 mm by bush hammering. All loose materials are brushed off and the base cleaned before application.

應用與底層的處理

抹灰是指在鋪砌建築材料前的底層作預備工作，塗抹混凝土或磚牆，使表面平滑後，才開始油漆、黏牆紙或其他室內裝飾工作。

批灰則泛指處理外牆的工作，準備鋪砌光面瓷磚、馬賽克，或噴漆、噴石粉等。

底層的處理是清除風化面、水泥乳、油漬、塵埃及其他疏鬆物質。在不同物質的底層面間，如混凝土、磚牆，須安置 150 毫米闊的網片，才可以進行抹灰工作。而處理垂直混凝土面或梁底，必須在板模拆去的四小時內甩灰塗底，砂漿以 1：2 至 1：2.5 的水泥與大粒的沙再混和水，使牆面凹凸不平，並要保持兩小時的濕潤，經堅固後，方可施加抹灰的底層。

現有的堅硬混凝土牆底部的翻新處理，可以直接用電錘鑽鉗混凝土面，至 3 毫米深，將鬆脫的物質清洗後，便可進行抹灰工作。

Lime/cement internal plaster

Internal lime plaster consisting of cement, lime, and sand in two coats on solid background is treated as follows:

1. The undercoat is not to exceed 10 mm thick to walls and 5 mm thick to soffits of 1:4:16 mix cement, lime and clean washed sand (by volume).
2. A finishing coat 5 mm thick of 1:12:30 cement, lime and sand.

A minimum of seven days is to elapse between the application of the undercoat and the finishing coat.

Disadvantages of lime plaster are drying shrinkage, which causes cracking and particles of sand left on the finishing surface.

Improvement in quality can be done by replacing the finishing coat by gypsum plaster with 25% of lime putty (formed by adding hydrated lime to water as a thick, creamy consistency undisturbed for 16 hours before use), which can give a smoother finish.

Another way is to add paper pulp (an elected size of baled paper mixed with water) at 16 kg per cubic metre of lime plaster. A smooth finish can be produced. However, an unsuitable mix can lead to cracks and small holes created on the plaster surface.

室內抹灰的石灰、水泥

室內用石灰抹灰含水泥和沙石，並用以兩層加在堅固的底層上，具體如下：

(1) 在牆身上的底層不能超過 10 毫米厚；在樑底的不能超過 5 毫米厚。以體積的成份比例為 1：4：16 的水泥、石灰及沙石。
(2) 完成的表面層為 5 毫米厚 1：12：30 的水泥、石灰及沙石。

從施工到完成表面層之間，最少須有七天的時間。石灰抹灰的缺點是乾涸後會收縮，引致表面破裂及留有沙粒。

第一種改良方法在完成的表面層抹上佔 25% 石灰膏的石膏灰（混合水與含水石灰，成為厚的、浮狀的形態，保存 16 小時後才使用），使表面更平滑。第二種方法是加入紙屑（經選擇的，再混合水）到每立米 16 千克的石灰泥漿中。混合物會較平滑，成為紙巾灰，但需留意不適當的混合過程會令石膏出現孔洞。

Ordinary Portland cement

Gypsum plaster

Gypsum is a crystalline combination of calcium and water. Gypsum as retarded hemi-hydrated gypsum plaster is used. A continuous smooth but hard surface ready for direct internal decoration can be produced. Application is by means of a wooden float or rule worked between dots or runs of plaster to give a true and level surface.

For general concrete and blockwork, two coats of plaster up to 10 mm thick are used:

1. An undercoat of 1:2 mix gypsum plaster and sand
2. A finishing coat of board finish plaster with up to 25% of lime putty added

White cement

Lime

Lime cement plaster

White cement with paper pulp

Metal lath to reinforce joints between different bases

If the background is a metal or a wood wool slab, a first coat of 1:1.5 mix plaster and sand with metal lathing is applied, forming three coats with a total thickness of 13 mm from the outer face of lathing.

Gypsum plaster can also be applied on gypsum plasterboard as one finishing coat 5 mm thick trowelled to a smooth surface using little water.

石膏灰漿

石膏灰是鈣與水的晶體型混合物。石膏灰是減低濕度的半濕石膏灰，可以做到持續平滑的堅硬面作室內裝飾表面。施工時以木條或間尺以抹灰點作標準量度厚度水平。一般的混凝土與磚牆工作，以兩層抹灰達至 10 毫米厚為標準：

(1) 底層為 1：2 石膏灰漿與沙石；
(2) 完成層為大面積的石膏灰，含高達 25% 的石灰膏。

如背板為金屬或木板層，第一層要為 1：1.5 石灰漿與沙石，再用金屬網，以 13 毫米總厚度作第三層抹灰處理。

石膏灰可用於石膏灰板之上並成為 5 毫米厚完成層。

Miscellaneous internal plaster

Renovation plasters offer better resistance to moisture and salts. A strong undercoat such as 1:1:6 cement, lime, and sand mix is applied followed by gypsum plaster as a finishing coat.

Projection plasters are specially formulated gypsum plasters suitable for direct application by special plastering machines. The projection plaster is applied by spraying, the whole process finished in a continuous operation to the required thickness. Increased productivity can be achieved.

X-ray-resisting plasters are premixed plasters incorporating barites (barium sulphate) aggregates in undercoat, and finishing coats.

Acoustic plasters are special plasters with suitable aggregates, e.g., pumice. They are premixed and applied to an appropriate undercoat plaster which is well keyed to give an adequate bond. Various texture finishes can be achieved.

其他室內抹灰工作

翻新性抹灰具有防潮與鹽份的作用。先塗上強勁的底層如 1：1：6 水泥、石灰及沙石的混合，再以石膏灰為完成的表面層。

噴射性抹灰是經由特別處理的石膏製成，可直接以機械完成。施工時以機器把抹灰平均噴射至指定的厚度，這方法可提高產量。

防 X 光抹灰是混合流酸鋇於底層與完成表層之間。

隔音性抹灰是特別的石灰，混合了適當的成份，如浮石等。不同的物質成份能做出不同質感的表面。

Plasterwork defects

Besides problems with materials or workmanship, plaster defects may be caused by structural background, climatic conditions prior to, during, and after the plastering process, and the correct choice of a plastering system.

Some defects are listed (adapted from BS 5492:1990) below:

1. Bond failure. This is due to relative movement between different plaster coats, weakness of the bond, residual moisture in the background, inadequate key on the background, insufficient time for setting of the plaster undercoat, or local exposure to heat, resulting in detachment of the plaster from the background (flaking), or blistering and shelling of the final coat.
2. Cracking. This is usually due to background structural movement or shrinkage movement in undercoats.
3. Crazing. Minor crazing is inevitable due to the nature of the plaster but can be better controlled by choice of well-graded sand, ganging lime/sand undercoats with gypsum plaster, observance of adequate time intervals between undercoats and final coats, avoidance of excessive rates of drying by exposure to strong sunlight, wind, and artificial heating.
4. Efflorescence. This is caused by the presence of soluble salts such as sulphates of sodium and magnesium in the background and adequate water to transmit these to the surface.
5. Grinning. This is generally caused by marked differences in suction of the background which are not masked by the usual procedure of adjusting the suction and which are manifested as areas of varying texture on the plaster surfaces. This can be the result of the use of mortars that are markedly different from the bricks or blocks.
6. Irregularity of surface texture. This can be caused by faulty workmanship but is acceptable under certain limits of tolerance.
7. Popping or blowing. This is caused by materials in the mix which expand after the plaster coat has set. Conical holes are left as 'pops' or 'blows'. This can be avoided by proper mixing of materials.
8. Recurrent surface dampness. A common cause is rain or water penetration for external walls or walls adjacent to sanitary fittings. Other causes may be the presence of hygroscopic salts attracting dampness when humidity is high.
9. Staining. Rust staining may be caused by application of unsuitable plaster or with impurities to, say, metal lathing or by the presence of sea salts in sand used in plastering which is in contact with metal work. Persistent damp conditions may cause rust staining on plasterwork in contact with corrodible ferrous metals.
10. Softness or chalkiness. This can result from excessive suction of the undercoat, undue thinness of the final coat, working past to setting point, or subsequent exposure to the final coat to excessive heat or draught during setting.

抹灰工程的缺失

除了物料與工藝等因素外,抹灰工作也會因結構、氣候、施工程序與方法等而有不同的缺失。

在 BS5492:1990 中列明以下缺失錯漏為:

(1) 結合損壞。由於不同灰泥層的活動、結合層的弱點、底層部份的介面充齒不足夠、底層含水份、灰漿未夠時間凝結或過份受熱,使表面的灰漿與底層分離。
(2) 裂縫。由底層部份的結構活動或底層收縮而產生的裂縫。
(3) 微絲裂縫。由石灰本身的特質所引致,因此可選擇較細微的沙石、處理過的石灰作底層,以改善石灰的情況。另在施加表面層時留有足夠的間隔,同時避免暴露於陽光、強風或熱力下而讓水份蒸發。
(4) 風化現象。底層含有水溶性的鹽如硫化鈉或鎂,過量的水份蒸發到表面時所引的現象。
(5) 磨面。通常由於底層的吸水力並非平均分佈所引致,這是因為在施工時,沒有使用任何方法來處理底層吸力,以及泥漿不平均所形成的紋理表面。
(6) 不平均的質感。這是工藝欠佳所造成,但少許失手是可以接受的。
(7) 凍裂或飛泄。灰漿中的微粒因膨脹而泄出表面層,形成很細小的孔洞裂縫。正確地混合物料可以避免此現象。
(8) 表面重覆潮濕。雨水由牆身滲入或附於牆身的渠管漏水,都會使表面不斷受潮。原因是水中藏有大量的鹽份,當濕度高時會吸引更多水氣到表面上。
(9) 斑漬。銹斑的形成,多是由於不適當的灰漿施工法,或由於本身含有大量雜質,又或沙中海鹽遇上金屬時所產生的化學反應。由於接觸長期的潮濕狀態而氧化金屬去產生斑漬。
(10) 軟化或粉狀化。這可能由於底層被過份地吸去水份、表面層過度減薄、施工時間超過凝結期,又或表面層暴露於過熱或過冷的環境。

Factors influencing the selection of a rendering mix

To select the right type of rendering, the designer should take into account the following (extracted from BS 5262:1976):

1. Choice of rendering mixes. This will depend upon the appearance desired, exposure conditions, nature of the background, and the functional requirements. These factors should therefore be considered together. Experience has shown that a porous rendering not stronger than the requirement needed for adequate durability and with an open- or rough-textured finish is likely to give the best results in the majority of circumstances.
2. Preparation of the background. The method used will be dictated by the type of background. Consideration will need to be given to the method of forming a key and to any materials to be used, including metal lathing. Consideration will also need to be given to the fixing of

any devices to be built into or attached to the background for support when adhesion cannot be relied upon.
3. Detailing of architectural features. Proper detailing of items such as copings, eaves and verges, string courses, damp-proof courses, flashings, plinths, and bellcasts have considerable bearing upon the durability of the rendering.

選擇抹灰成份的不同因素

選擇正確成份的抹灰因素如下：
(節錄於 BS5262：1976)：

(1) 抹灰的成份比例。主要視乎外觀的要求、處身的空間環境、底層的成份和實際的功用，這幾個因素都需一併考慮。據一般的經驗，含不超過本身所需強度的滲透性抹灰料，並有粗糙的表面，會較容易做出完美的效果。
(2) 底層的預備工作。施工方法受底層的類別和成份所限，但最重要是預留齒痕為抹灰時相扣的面，有時亦會以鐵網為底層的齒痕。或者當有附加之東西需要掛上時而不能用膠漿，亦應一併考慮。
(3) 建築的裝飾細節。合適的細部設計，如蓋頂、屋檐、絃帶、腰線、防層、擋水板、柱腳等，可以使抹灰更加耐用和持久。

Lime/cement external render

External rendering of cement, lime, and sand is applied in two coats with total thickness not exceeding 20 mm:

1. Undercoat of 1:2:6 mix (by volume) of 10 mm minimum thickness. This is applied from the float, trued off, scratched, and allowed to set for a minimum of three days.
2. Finishing coat of 1:3:6 mix of 10 mm maximum thickness and textured finish with the float. This is a weaker mix but will reduce shrinkage in the finishing coat.

Cement, lime, and sand mixes have a lower drying shrinkage but are more absorbent than are cement sand mixes. However, after periods of rain, these will dry out rapidly.

石灰/水泥的外部抹灰工作

戶外應用的抹灰工作以水泥、石灰和沙石混合為抹灰料，兩層的總厚度不能超過 20 毫米：

(1) 以 10 毫米為主的 1：2：6（容量）的比例，厚度最少要有 10 毫米。應用時以浮面或齒痕等形態表現，凝固時間須有三天。
(2) 完成的表面層以 1：3：6 混合，厚度最多為 10 毫米，完成後具豐富的質感，此混合料雖較薄弱，但可以減低收縮的可能性。

水泥、石灰和沙石混合灰具很低的乾涸收縮性，但較水泥沙石混合成份更具吸水性能。不過，經連場大雨沖刷下，也會很快地乾涸。

External renders to receive tiling

Various cement renders

Cement render

Cement and sand rendering are strongly moisture resistant but subject to cracking due to high drying shrinkage.

A ratio of 1:3 cement sand mix is used. For rendering not exceeding 10 mm thick, one coat is sufficient. For rendering exceeding 10 mm thick, two separate coats are applied.

Waterproof admixture can be added to the cement sand rendering for use in cement dadoes in kitchens, bathrooms, balconies and open staircases, internal walls of closets and water tanks, and other areas to be waterproofed.

水泥抹灰

水泥與沙的抹灰料，有強烈的防水性質，但因其快乾特性而容易引致龜裂情況。

普通使用的是 1：3 的水泥沙混合。如抹灰不厚於 10 毫米，單一層的抹灰料便已足夠；如厚度超過 10 毫米，就需要兩層的抹灰。

如用於廚房、浴室、露台、露天梯間等牆裙地方、水缸以及任何需要應用防水的地方時，可加入混合防水劑。

Exposed aggregate rendering or 'Shanghai' plaster

Shanghai plaster gives a total thickness of 20 mm in two coats:

1. An undercoat 10 mm thick of 1:3 cement and sand mix
2. A finishing coat 10 mm thick of 1:1 cement and stone chippings which are selected in different proportions of colour chippings to produce dark grey to white colour. Before setting the finishing coat, the surface is scrubbed to expose the aggregate.

外露的粒狀灰或「上海」抹灰

(1) 先以 1：3 的水泥沙混合做 10 毫米厚的底層；
(2) 完成的層以 1：1 的水泥、碎石混合，再做 10 毫米厚度。隨著碎石顏色的不同，可以做出不同的顏色變化，由灰至白等。如表面層還未凝固，再以掃刷把細粒料表露出來。

Floor screeds

Screeds are laid on concrete floors and can be self-finished or prepared to receive finishes such as flexible tiles, carpet, ceramic tiles, asphalt, timber block flooring, or marble.

Generally a 1:3 cement sand mix with minimum water for workability is used. To obtain a good key for the concrete surface, any mortar droppings and laitance scrum are removed by brushing, scrubbing, and hacking. The surface is then thoroughly wetted to minimize absorption. A thin coat of cement slurry is also added before the screed application. The screeds are laid in bays of 15 m² maximum and in a pattern. At least 24 hours are allowed between laying of adjoining bays.

Usually, a minimum 20 mm thick screed is used to bond to a hardened concrete base. For screed not bonded to a base or floating screed, 50 mm to 65 mm thick screed including tile finish is used. For floor screeds thicker than 40 mm, a mix of 1:1.5:3 cement, sand, and coarse aggregate is used. Steel reinforcement can be added for further strength.

Lightweight screeds of 1:8 cement and lightweight aggregate, e.g., 5 mm exfoliated vermiculite, can be used on roofs with 50 mm minimum thickness, affording a certain degree of insulation. A lightweight screed of 1:6 mix is used for floors.

The surface of screed can be floated with a steel trowel to provide a smooth and level finish as a self-finish or for further laying of finishes like flexible tile flooring. Other self-finished surfaces can be by wood float to obtain an even-textured surface or by stiff brush to give a roughened texture. A surface hardener can be applied on the screed, usually while green, to obtain a hard-wearing finish suitable for car parks.

地台抹灰

地台抹灰的作用，是使混凝土地台成為完成的表面，或作為鋪砌地台料之前的打底，然後可鋪砌地台磚、地毯、瀝青、木地板或雲石等。

通常以 1：3 水泥沙石混合最少水量來應付一般的工作。為達到良好的混凝土接口齒，所有的硬水泥粒等雜物須掃淨，底面需稍濕來減低施工時水泥的吸水力。在水泥還未刮板前，要先塗一層薄薄的水泥泥漿為基礎，隨後刮板。地台抹灰以最多 15 平方米的面積，分格進行施工，格間要預留 24 小時的間隔。

一般來說，用 24 毫米厚的刮板來處理較堅硬的混凝土底，而用於浮面的刮板時，則連同完成鋪面料設計，需有最少 50–65 毫米的厚度。當地台的刮板有 40 毫米厚時，需 1：1.5：3 水泥、細沙、粗粒粒料混合，同時要加入鋼筋作強化用。

輕抹灰可用 1：8 水泥與輕粒料，如 5 毫米的碎輕粒料作混合，可用於屋頂，能提供隔熱的功能，但最少要有 50 毫米厚。而 1：6 的比例成份則可用於地台。

抹灰表面可以用鋼鏝刀刮平,，使完成之後能光滑平坦，亦可鋪砌表面用彈性地台。另一方法，是以木條刮平，再用硬掃刷出不同的質感。表面凝固劑可加於未凝固的刮板料之中，通常用於停車場等較需要更堅硬的地台上。

Laying of floor screed

2.14 Painting
油漆工程

Painting

Paint is a mixture of a liquid or medium with a colouring or pigment. The application of paint to the building elements provides a protective coating and to impart colour.

Paint is applied in several coats.

油 漆

油漆是一種混合了顏料色素的液體，其作用為保護性的表面層，兼具美觀的效果。

油漆工作需要連續幾個塗層。最基礎的底層作為完成建築物的表面，亦可保護底層，避免與外層油漆產生化學反應。第二層的底油作為完成表面層的底部處理。最後的完成表面層可有許多不同顏色選擇，一般的表層油漆都含有人造纖維素，可以減低油層的乾涸時間，亦使油漆較易塗上。不過，塗油漆時必須依據供應商的說明標籤為準。

Application of paint and the layering
油漆的應用範圍及不同塗層

Base surface 底層材料	Finishing paint 完成表面油	Primer 底層的油
Internal and external woodwork 用於室外、室內的木工	Synthetic finishing paint 人造完成油	Aluminium primer 含鋁質的底油
Internal and external metalwork 用於室外、室內的五金工程	Synthetic finishing paint 人造完成油	Zinc chromate primer, metallic zinc-rich primer, or lead primer 鋅化鉻、鋅或鉻
Galvanized metal 含鉛水鐵器	Synthetic or non-toxic paints 人造或不含毒	Etching primer with a zinc chromate base, calcium plumbate 鋅化鋁底
Internal and external metalwork 用於室外、室內的五金工程	Polyurethane paint 聚合尿酸油	Polyurethane red lead primer 聚合尿酸性紅鋁底
Cement lime-plastered surfaces 水泥、抹灰的表面	Emulsion paint 乳膠漆	Prime lime 石灰底

Diameter of pipe 喉管的直徑	Minimum fall 最少的斜度標準
100 mm	1 in 40
150 mm	1 in 70
225 mm	1 in 100
300 mm	1 in 150

Primers
The priming coat or primer is the first coat of paint used to seal the surface of the building element, acting as a barrier to prevent any chemical action between the surface and the finishing coats. It is also applied to enhance adhesion, prevent absorption of later coats by porous surfaces, and give corrosion resistance over metals.

Sealers
Sealers are clear or pigmented materials applied in thin coats to prevent the migration of substances from the substrate into later coats. Sealers are usually used on top of cement plaster or concrete.

Undercoat
The undercoat is used to build a protective coating, which provides a good surface for the finishing coat. It should also provide a fresh surface of uniform texture and a colour close to that of the finishing coat, except colours like bright red, yellow, or orange, to which a white colour undercoat will be applied to bring out the brightness of the finishing colours.

Finishing paint
Finishing paint comes in a variety of colours and finishes and usually contains a synthetic resin for quick drying and ease of application. All application of paints should be in accordance with the manufacturer's data sheets.

Textured finish
A textured finish gives a distinct and attractive three-dimensional appearance for internal and external walls. Nowadays, the application of the textured finish can be achieved in different ways. One common application method is by rollers or spraying machine. Various texture renderings like mountainous, head-cut, or some special design is popular on cementitious surfaces and masonry.

Components for a painter's work

Emulsion paint on plaster

Stains for a timber door

以下是常見的油漆層：

油漆底層
底層用來增加黏附力，防止之後的塗層與基層物質起化學作用或吸收其物質，並且給金屬增加抗腐蝕性。

密封塗料
密封塗料是透明或者有顏色的薄塗層，用以防止有物質從底層移至之後的塗層。通常直接用在水泥層或混凝土上。

底層油漆
底層油漆提供一個有紋理的新表面以及一種與完成油漆相近的顏色。至於紅色、黃色和橙色的光亮表層，可用白色底層油漆提升亮度。

表層油漆
作為最後的完成面，表面油漆有不同的顏色及質感供選擇。表漆成份通常為合成樹脂，既可減低油層乾涸的時間，也使油漆較易塗上。不過，油漆程序必須按照供應商說明標籤的描述為準。

紋理表面
紋理表面給室內和室外的牆帶來了特別和引人注目的三維外觀。如今，紋理表面的效果可以用不同方法達成，其中最常用的一種方法是使用滾軸和噴霧器。不同的紋理如山紋、切紋或一些特別的設計在水泥表面和石面都很流行。

Types of paint for different surfaces

There is a lot of variety of paint, the choice of which depends on the use of the building element and the aesthetics desired.

On plaster, render, concrete, brick, or block, emulsion paint with a large variety of colours is commonly specified. For the same surface, anti-mould acrylic emulsion paint is used in areas with high humidity. Lime wash and cement paint are applied to service rooms. Epoxy paint with or without texture can be applied for both internal and external walls with water repellent effect.

On concrete roads, tar macadam, and similar surfaces, road-marking paint is used.

On structural steel surfaces, iron oxide paint with a suitable primer is used to resist rust.

On metal surfaces, synthetic paint and metallic paint are commonly used. There is also heat-resistant paint and acid-resistant paint for special application. Enamel coating, polyurethane, or epoxy paint can be very durable even in external conditions. Modern developments of fluorocarbon coatings or fluropolymer resin-based finishes are effective protection for curtain walls, column covers, signs, spandrels, etc.

On wood surfaces, lacquer paint and stains are common for fine finishes. Wood preservative can also be introduced for additional protection. Synthetic paint, polyurethane paint, and varnish can give a tough finish.

Newly invented paint for eco-friendly reasons

It has been proven that volatile organic compounds (VOC) in ordinary paint and adhesive will cause allergic symptoms in people, even a small amount in the air.

Regulations were established in Hong Kong to reduce the amount of VOC in paint. Therefore, paint producers have invented different formulas for more eco-friendly paints. Some paints have innovative polymer synthesis to eliminate paint odours, low VOC, and a certain level of harmful formaldehyde in interior spaces.

油漆類別與各種物料的關係

油漆的種類有很多，主要視乎建築物的需要與對美感的要求去作選擇。水泥、抹灰、混凝土、磚塊的表面，常採用不同顏色的乳膠漆。如環境比較潮濕，可以塗上防霉人造乳膠漆。水泥灰底的油漆較常用於一般房間。平滑面或粗糙面的環氧性油漆可用在室內及室外的牆身，兼具防水功能。

亦有專門塗於混凝土或柏油路的道路標記油漆。

在結構用的鋼材亦有防銹功能的油漆。在金屬表面上，則用人造油漆與金屬油漆為主。當中亦可以找到隔熱的、防酸的油漆。磁油表層、聚氨酯或環氧性油漆，都是比較持久而耐用的，即使是用於戶外的環境。隨著科技的發展，含碳氟化或氟化高聚脂的鍍面處理，可有效保護各種金屬建材，如玻璃幕牆、覆蓋板等。

至於木材表面的處理方法，尤其是微細的木器表面，主要是以油漆或染色方法來處理。可同時加入木材保護劑。人造油漆、聚合尿酸油漆都可以做出不同程度的耐用表面。

新發明的環保油漆

普通油漆裏的揮發性有機化合物,即使很微量,已證實會給人類帶來過敏症狀。

香港近年已建立有關減少油漆中揮發性有機化合物含量的法規。因此,油漆生產商研製了不同的配方,提供對環境更友善的油漆。有些油漆加入了革新的聚合物,用來消除油漆中的有害成份、低揮發性有機化合物和滿佈於室內的有害甲醛。

Workmanship

Surfaces adjacent to paintwork areas should be properly protected before application of paint, e.g., weather stripping on metal windows or doors. All holes, cracks, and defects in the surfaces are repaired before painting. Paint is applied in even thickness by quality bristle brushes of suitable size. If specified, mechanical spraying machines are used for spray coating, which should be in accordance with the manufacturer's instructions. Sample panels are usually painted before commencement of work.

施工技術

需要油漆處理的表面必須經過小心細緻的程序。先作好油漆旁邊的底層保護,如金屬窗、門的防水帶;表面的孔洞、裂縫等要先填平處理才可以塗上油漆。施工時,要以平均的厚度與速度進行。如有需要,可在合同條款中列明要以機器噴嘴作勻稱的塗油工作。工作前要做樣板以控制品質。

2.15
Builders' Work in Relation to Plumbing, Drainage, and Mechanical and Electrical Services
有關渠道、水喉、機電工程的建築項目

Drainage and plumbing regulations

The installation of drainage and plumbing work is governed by the Building (Standards of Sanitary Fitments, Plumbing, Drainage Works and Latrines) Regulations, including items for standards of sanitary fitments, plumbing—soil fitments and waste fitments, plumbing—pipes and eaves gutters, drainage work, septic tanks, cesspools, testing of drainage work, certain work to be carried out by the Building Authority, latrines, etc.

To establish the minimum number of water closet fitments, urinals, lavatory basins, baths or showers required, the usable floor space (which means the total floor area of a building excluding public circulation area, building service rooms and toilets, kitchens, and bathrooms) is calculated. There is a provision rate of persons for different uses of the building, e.g., one person every 3.25 m^2 for tenement houses, one person every 9 m^2 for offices and industrial buildings, one person every 1.5 m for restaurants. The number of persons in places of public entertainment and cinemas is based on the maximum capacity of the area. The number of persons living in a residential building is determined by the Building Authority. After the number of male and female persons is calculated, the number of fittings can be established from specified tables in the Regulations.

For water closet fitments, the flushing cistern discharges water on each occasion not less than 9 litres and not more than 14 litres. The soil fitment is provided with a suitable trap of water seal not less than 50 mm. The internal diameter of such a trap is not less than 80 mm but can be modified to 50 mm upon approval of the Building Authority.

Water closet cubicles for persons with disabilities are not less than 1.5 m × 1.75 m in area. A water closet and a washbasin suitable for use by people with disabilities are installed. These are also at least three grab bars fixed on the walls. The door has 750 mm clearance and opens outwards.

Ventilating pipes are provided for drains and sewers projecting not less than 1 m above the roof of the building where they are attached.

Drainpipes, anti-syphonage pipes, and ventilation pipes inside a building can be fixed inside a pipe duct which is provided with access panels to reach all pipe connections.

Falls of drains are also noted, as these may cross with structures or affect headroom.

Utility holes are provided at the change of direction of drains (except small changes) and not exceeding 60 m in length. Covers for utility holes are standard cast iron double-sealed covers. A second cover incorporating the adjacent floor finish can be added for aesthetics.

Cleaning eyes are provided at bends for soil pipes and waste pipes, to facilitate cleaning. Thus, access panels are designed for accessibility to these cleaning eyes, e.g., at the side or end wall of a bathtub, a false ceiling to reach the cleaning eyes of the upper floor.

Cesspools, septic tanks, and sewage treatment plants are constructed in accordance with the Regulations. Testing is carried out to satisfy the Environmental Protection Unit. Grease traps are provided for commercial kitchens and car parks, to reduce contamination.

Connection of the drain to public sewers and construction of the last utility hole are carried out by the Drainage Division of Highways Department. Such costs involved will be recovered from the owner of the development.

An aperture for a gas water heater is required to be provided in bathrooms and shower rooms. The size of the standard aperture is 380 × 380 mm or 240 × 240 mm when only a shower is installed. The location of the aperture is governed by regulation with an unobstructed area around it. If no heater is fitted during construction, the standard aperture can be sealed by easily removable brickwork or blockwork. However, the position is made distinguishable.

渠道與水喉的建築條例

安裝水喉、渠道等事項都是依據政府的建築條例（衛生設備標準、水管裝置、排水工程及廁所）為標準設計。項目包括制定標準的衛生設備、供水系統、污水及廢水管、雨水管、簷溝、排水工程、沙井、化糞池、喉管測試、某些由政府負責的工程、廁所等。

在計算最低要求的衛生設備數量時，必須以實用面積（指所有的建築面積，但扣去公眾地方通道、公共設施的房間、浴室、廚房等）計算。而各類型的建築物房間都會指明某特定系數為標準，如租用的舊式樓宇以每 3.25 平方米為一個人的標準空間系數；在辦公室或工業用途的地方以 9 平方米空間設計為標準；酒樓用途的，以 1.5 平方米為標準。至於娛樂場所和戲院，則以該面積可容納最多人的數目為計算標準，以住宅用途平面，則由建築事務監督自由決定。決定了男女人數的比例後，便可以計算設計的標準了。

政府條例規定，在連接坐廁系統時，每次的沖水量必須保持在 9 至 14 公升之間；便具等存水彎喉管必須存水 50 毫米以上。去水喉管直徑應大於 80 毫米，但亦可向建築事務監督申請批准改為 50 毫米。

傷殘人士所用廁所潔具必須在 1.5 × 1.75 米以上，同時配置專為傷殘人士設計的洗面盆及水龍頭等；在牆上安裝三條不等高的扶手，另外門必須有 750 毫米濶及向外推出。

每條去水及污水管都要設計一個通氣管道，此管道出口要距離屋頂的完成面最小 1,000 毫米的高度。

建築物內所有的去水管、反虹吸管、通氣管等，必須安裝在指定的喉管空間中，而且還要特別裝設供修理保養用的維修門。

喉管設計時因要符合足夠的斜度，會引致建築物可使用的空間高度減低或與結構相撞，這些問題都要留意。

接駁轉向喉管（轉變方向較小的除外），或當喉管超過 60 米長時，都要多設計一個沙井於喉管中間。沙井面蓋是由標準的生鐵製造，具有雙重關閉的介面。如設計偏重美觀，會再附加裝飾面蓋以達到視學的美感。

Bathtub installation

Drainpipe for water closet

在污水管的轉彎位置都裝有「清潔眼」，方便清潔和修理。在設計討蔽式喉管時，必須留意「清潔生口」的位置，如在浴缸的間牆、假天花等，並使修理門能順利達到維修的用途。

化糞池及污水處理井需根據建築事務監督條例建造，並且要符合環境保護署的測試標準。在一般商用的廚房、停車場等，更要建造去油井，以避免油污排到公共渠務設施中去。

一切接駁污水至公共沙井或喉管的工作，都是由渠務部或路政部執行的，而駁渠的費用則由業主或發展商負責。

在浴室中，要預留 380×380 毫米或 240×240 毫米的孔洞，為日後加建氣體熱水爐的設施裝置，此孔洞的標準必須依據建築條例的限制與標準。施工時未必會安裝熱水爐，此孔洞可用磚頭或其他可移動的材料填塞，但位置必須要容易被發現。

Plumbing pipework

For the city water supply system, materials for pipework include galvanized steel, copper, and unplasticized PVC. Copper pipes or plastic-coated copper pipes are used for hot water systems and inside the bathrooms. Copper pipes are provided with expansion joints over 12 m long. However, care must be taken at installation to avoid any damage which may lead to subsequent troublesome water leakage when all finishes are installed. Ductile iron pipes are used to stand heavy loads when the system of pipework passes through carriageways.

For a flush water supply system, ductile iron pipes are used as delivery pipes for salt water; galvanized steel pipes are used as delivery pipes for city water, nullah intake, or well. Other pipes of the distribution system are PVC pipes, which are kept a minimum distance of 150 mm from hot water pipes.

Pipes are run in walls, floor screeds, or pipe ducts. Pipes through walls and floors are first cast in PVC sleeves with 2 mm to 12 mm clearance for expansion and movement. Mastic sealant can be used to achieve watertightness. If through fire-rated walls and floors, galvanized mild steel pipe sleeve with 20 mm clearance is used. The clearance space is then caulked with mineral wool. Pipes through flat roofs are by cast iron or galvanized mild steel sleeve with 2 mm to 12 mm clearance projecting 150 mm above the roof finish. Caulking is then applied by mastic sealant. Pipes laid on the roof are bedded on concrete blocks at approximately 1 m centres and 75 mm clear of roof finishes.

管道工程

城市自來水喉管的材料是鉛水鐵、銅、非軟性塑膠。在浴室內或熱水系統所用的一切喉管，都以銅管或包膠銅管製造。銅管長度如超過 12 米，須在中間設計伸縮縫，且在安裝時必須特別小心避免漏水。如喉管需穿過馬路，就要採用具伸縮性的鐵管，以承擔載重負荷。

沖廁用鹽水是以具韌性的鐵管傳送，而城市用水則以鉛水鐵管傳送。其他塑膠喉管距離熱水管最少須 150 毫米。

喉管安裝在牆身、地台批盪或管道空間。所有牆身及地台喉管都會以塑膠套通封好，並以 2 至 12 毫米的空間作為伸縮縫，容許喉管作輕微的活動。膠脂劑可用來保持喉道的潔淨。喉管經過防火牆和地台位置時，必須使用鉛水生鐵管通並侏持 20 毫米隔離，而這空隔則要填滿纖維棉。至於喉管經過屋頂時，以生鐵或鉛水生鋼套通封住，之間留 2 至

12 毫米的距離，且須保持距離屋頂完成面 150 毫米的高度。堵縫用瀝青密封膠填塞。屋頂上平放的喉管須以混凝土磚承托，以 1 米間的中至中距離和 75 毫米高度距離地面安裝。

Drainage pipework

Materials for drainpipes include concrete, vitrified clay ware, cast iron with flexible joint, ductile iron, and unplasticized PVC. PVC is used inside buildings. For down pipes, cast iron is commonly used. Concrete and vitrified clay ware can be constructed in large-diameter pipes. Ductile iron is used when crossing carriageways with heavy loads.

Laying of pipes outside the building is first done by setting out with levels. Excavation is done in trenches. On completion of the excavation, the pipes are laid. Beds and surrounds are provided to support the pipes:

1. Natural bedding. For places like planters and lawns, where light traffic is envisaged and for short lengths of pipe, natural bedding by trimming the earth to provide uniform and solid bearing is acceptable.
2. Granular bedding. Bedding with coarse aggregate provides some support to the pipe which is applicable on virgin land and light traffic.
3. Concrete bedding. Pipes are laid in continuous concrete beds or precast cradles and then the haunch and surround after testing. In areas like newly filled land where settlement is possible, short steel piles called monkey piles are provided for the foundation support to give the strongest bedding for pipes.

After bedding is complete (for concrete bedding, wait for 48 hours), the trenches are backfilled and compacted.

Sub-soil drainpipes can be concrete porous drainpipes, unglazed clay ware field drainpipes, perforated vitrified clay ware pipes, or perforated plastic pipes. Natural bedding or granular bedding is provided.

Overflow pipes for roofs are constructed with 75 mm PVC pipes projecting 75 mm over the side of the external wall. This design is to avoid water coming down into the building in case the surface channels are blocked up.

Laying of concrete pipes

排水管道

用於排水管的物質包括混凝土陶瓷結合瓦、具伸縮駁口的生鐵、韌性鐵與未增塑的塑膠等。塑膠可用於建築物內部。而排出雨水的垂直管多以生鐵為主。韌性鐵管則用在車路下面。

在安裝建築物外面的喉管時，先在路上掘一道深壕，才放置喉管，喉管的兩側以不同的物料填滿來保護和支撐喉管：

天然底層：在花槽或草地裏的喉管，因上面的負荷力不會太重，所以喉管只須放置在天然泥土面便足夠了。

粒料底層：喉管鋪在普通或承受車輛負荷等路面時，要以較大粒料作為喉管的底面層，以支撐喉管的安裝。

混凝土底層：喉管鋪在混凝土的底層或磚支架上，底層的混凝土經水管測試後，再澆灌混凝土去填塞。此方法用於較新且有沉陷機會的土地。短的鋼鐵樁（monkey piles）使喉管得到最堅固的支撐。當完成好底層之後（以混凝土的方法，需待 48 小時後），用泥土填滿壕溝。

Drainpipe for washbasin

Water supply pipes on roof

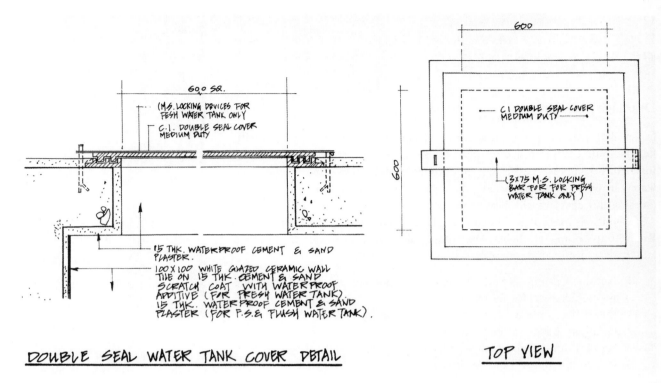

Double-sealed water tank cover details

泥面下排水管以半滲透性混凝土、無釉黏土排水喉管、疏孔黏土或膠管等為材料。一般以天然或粒料層為鋪底方法。

屋頂面的溢出喉管，要以 75 毫米直徑的塑膠管建造，另須離開外牆 75 毫米的距離。這個設計是當明渠被阻塞時用作疏導屋頂滿溢的雨水。

Fire service installation related work

Fire service inlets and sprinkler inlets poured in steel boxes with label panels are located near the entrance of a building at a prominent location for ease of access in case of fire. For a better appearance, a concealed type of design is often used for the panel doors.

Hose reels are often concealed in recessed housing in a typical floor. The control valve and nozzle are at a position not more than 1,350 mm from the finished floor level. Consideration has to be made to ensure that, when the closet doors are open, the width of the escape route is not decreased. Alongside hose reels are fire alarms and break glass call points which can be installed as recessed fittings also. The call points are not more than 1,200 mm above the finished floor level.

The hydrant outlet and riser are usually installed at the corner of the staircase landing without affecting the swing of the staircase radius. Outlets are at height of not less than 800 mm or more than 1,200 mm above the finished floor level.

The sprinkler installation affects the design of the ceiling and requires coordination with the structure to allow for sleeve openings.

Sprinkler tanks can be located underground or on the roof, depending on the structure, but these occupy a lot of space. A twin-end feed tank is smaller

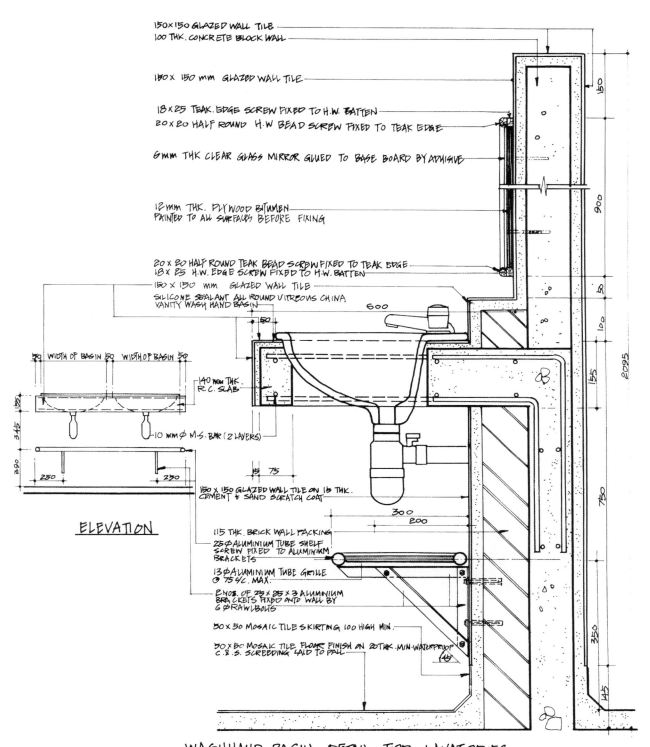

Washbasin details for lavatories in a rehabilitation centre

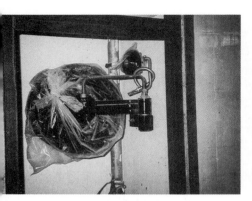

Hose reel installation

Smoke vent for basement

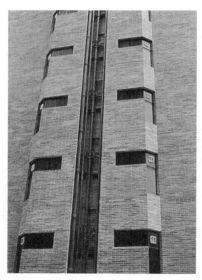

Drainpipes at the valve external wall recess level

than a single-end feed tank. Fire service tanks for the hydrant system are located on the roof, to provide static pressure.

For a large development, a fire control room with annunciation panel of fire alarm locations is required. This is located near the entrance of a building.

Signs relating to fire services are required. Notices stating fire doors are to be kept closed are required on both sides of fire doors, such as for smoke lobbies. Exit signs with built-in lights are required to indicate the exit door locations.

Some of the service rooms are classified as special hazard areas requiring enclosure by non-combustible construction having an FRP of not less than two hours, four hours for an adjoining staircase, and any doors provided with FRP not less than one hour. These include: high-voltage electrical switch gears, transformers, fire service pumps, air handling unit plant, air-conditioning unit, lift and escalator machines, dangerous goods store, and boilers.

有關消防措施系統

建築物入口地方的附近，必須設置消防入水制與花灑入水制，並要以獨立的鋼箱裝載，還須附有註明用途的門，方便火警時使用。如要求美觀，可用隱蔽式設計取代玻璃門。

消防喉轆通常設計在每層樓的暗槽。據消防條例列明，滅火喉噴嘴和控制閥必須裝在距離地面不超過 1,350 毫米高的範圍。在開啟置放喉轆地方的門時，亦必須留意消防逃生通道有沒有受阻。在消防喉轆鄰近，必須加設消防警鐘和擊碎火警鐘掣，而且規定要裝在距離地面不超過 1,200 毫米高的範圍。

至於消防入水口，一般都是設在樓梯的轉角位置，盡量保持樓梯位中 90° 弧形的闊度標準。而放水泊管道口距離地面必須在 800 至 1,200 毫米之間。

因建築結構的設計，花灑水缸可設在地庫或天面之上，體積相當大。雙頭的回水缸比單頭的為小。而消防水缸則規定要設在天面之上，使喉管保持自然穩定的水壓。

在較大型的建築計劃中，須在建築物入口的當眼地方，建造由一組控制人員負責的消防控制室。消防設施都需有適當的圖示火警說明。防煙門前後要掛上「保持關閉」告示，而走火門須有出路燈箱指引路向。

有些設施房間被列為特別危險地方時，必須以防火物料作為建築牆壁，一般阻隔防火時間為兩小時；或是以四小時隔開樓梯，並所有的門都須一小時的防火時間為標準。屬危險類別的房間包括：高壓電制房、變壓房、消防泵房、空調機房、電梯及升降機房、危險倉庫和鍋爐房等。

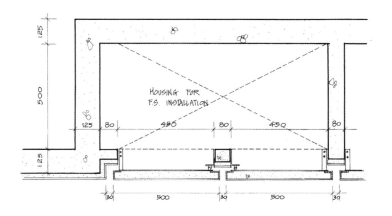

PLAN FOR F.S. INLET & SPRINKLER INLET

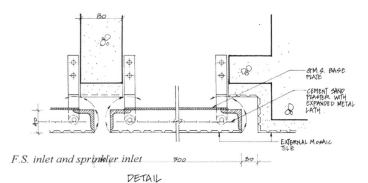

F.S. inlet and sprinkler inlet DETAIL

F.S. inlet and sprinkler inlet

Hose reel section

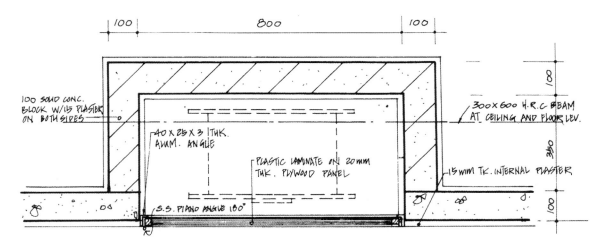

FS inlet, sprinkler inlet, and hose reel installation

Sprinker and FS inlet

Exit sign

Wiring work

Electrical-related work

The transformer room or substation has the bulk of electrical work. A route for transportation of equipment to the room (usually at ground level) is not less than 2.0 m wide with a clear headroom of 2.6 m. The transformer room has a clear headroom of 3.45 m for accommodation of equipment. An external wall is required for access and ventilation. A switch room of adequate size is located adjacent to the transformer room. The Hong Kong Electric Company and the China Light and Power Company have their own requirements for the transformer room, such as construction of the doors.

Meter rooms and electrical/telephone ducts located in typical floors are fitted with fire-rated doors, screwed on, or self-closing. Exposed trunking in the ceiling of a smoke lobby can be protected by a fire-rated ceiling like durasteel.

機電相關工作

變壓房或分站是電機工程中體積最大的項目。在工程計劃中，須考慮統籌所有機械設備的運輸路線，使龐大的機械可直接運送到機房（通常設在地面）。通道的設計，要有至少 2 米闊和 2.6 米高。變壓房的基本設計，須有 3.4 米的高度，且外牆的設計要能進出維修及通爽。適當體積的電掣房須建在變壓房附近。不同的電力公司，如香港電燈有限公司、中華電力有限公司，都有各自不同的基本規限，如對變壓房門的要求。

每層樓的電錶或電掣／電話喉管，都是安置在一個特別的房間，裝有防火門及自動關閉門鎖。至於在電梯間外露的喉管則以防火鋼鐵板包裹。

Lift-related work

The design of lifts is subject to the Lifts and Escalators Ordinance and approval by the Electrical and Mechanical Services Department. Lift shafts and lift pits are constructed with at least two hours FRP, i.e., 125 mm minimum thick reinforced concrete. Where there is accessible floor space below the lift pit, additional structural requirements for the lift pit floor may be imposed by the Building Authority.

Machine rooms have clear heights not less than 2.1 m. The doors of such rooms are self-closing and open outward, with an FRP not less than half an hour.

For lifts that do not have landing doors in some floors, emergency access to the lift shaft is provided at intervals not greater than 11 m, for the sake of evacuating passengers in the lifts during an emergency. Such access has a minimum height of 1.8 m and a minimum width of 0.5 m, opens outwards from the lift/shaft, and has an FRP not less than one hour. A sign marking the rescue door is displayed.

Every landing door is constructed with not less than a one-hour FRP. Adjacent to the landing door is a sign, 'When there is a fire, do not use the lift', which is incorporated as part of the lift lobby design.

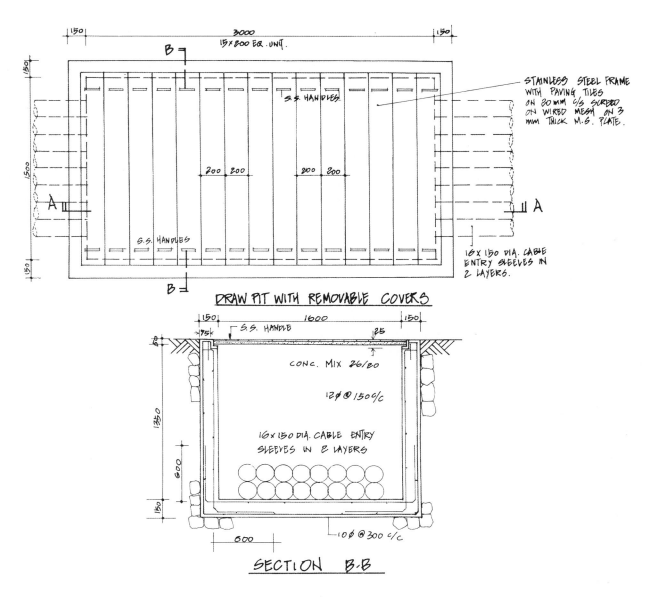
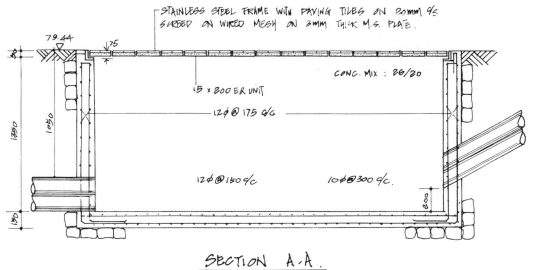

Cable drawpit for entry of cables to the transformer room

Lift door with textured steel finish

Lift interior with embossed steel panel

Engraved lift door (stainless steel)

升降機相關工程

升降機的設計必須依據政府機電工程署有關的《升降機及自動梯條例》及該署的指引。升降機的結構間與底槽須具兩小時的防火建築設計，即 125 毫米的厚度。如在升降機底槽之下仍有可用的空間，建築事務監督或許會附加設計的要求。

升降機機房必須要有最少 21 米的高度，門的設計要自動關閉、向外推出以及需有半小時的防火效能。

如果升降機並不是停於每個樓層，便應在 11 米距離間設計可供逃生的緊急活動門，此門最少要高 1.8 米及闊 0.5 米；門向外推出，並有一小時的防火效能，在此逃生通道間更要安裝足夠的照明設施。升降機內每扇門都要有一小時的防火設計，在升降機大堂的當眼位置，要掛上「如遇火警，切勿使用」的字句。

Air-conditioning–related work

Central air-conditioning plants are exempted from gross floor area calculation, but they still occupy a lot of space and generate noise and vibration, thus requiring acoustical consideration. Chilled water pipes, air ducts, and air handling units occupy space within typical floors. Louvres are required on the exterior of buildings for intake and exhaust air.

For window-type air-conditioning units in residential buildings, wall boxes or platforms are constructed as amenity features to improve the structural security of these installations. Noise, vibration, exhaust air, or dripping water are also considered.

冷氣相關工程

中央冷氣機房是可以豁免計算在建築面積內，因為它所佔據空間的面積並不少，更會產生大量的噪音與震盪，在設計時要考慮隔音問題。樓面間必須預留足夠的空間來容納雪種喉與水喉等。機房外牆須有百葉窗作排氣及抽氣之用。

一般住宅冷氣系統以獨立窗口機為主，只須在外牆窗口位置設計混凝土底板，承托露於外間的機身；安裝時，除了顧及結構穩固外，還要考慮噪音、震盪、排氣、去水等問題。

Refuse-collection work

Refuse collection is governed by the Building (Refuse Storage Chambers and Chutes) Regulations.

The refuse storage chamber is provided for domestic buildings with more than 1,320 m² usable floor area and for non-domestic buildings with more than 3,960 m². Car parks, industrial buildings, schools, and churches are exempted. For very large buildings, the storage chamber is designed to permit vehicular access for removal of refuse containers.

The storage chamber is located so as to have at least one external wall. Except for the access door and the hopper, there are no openings inside the chamber and no dimension less than 1.5 m and a ceiling height not less than 2 m. Construction is usually of brickwork, blockwork, or concrete. The internal finish for walls is glazed tiles. For ceilings it is smooth cement rendering, and for floors it is quarry tiles laid to fall towards an outlet drain and with coved tile at the floor and wall junction. A water supply point is installed inside.

The access door to the storage chamber is a close-fitting steel door without any internal projections. The size is not less than 1.8 m high × 1.25 m wide.

The refuse hopper is situated in the upper floors in a place with permanent ventilation to the open air. Construction is of stainless or galvanized mild steel plate at least 3 mm thick, the mouth of which has a clear opening between 250 × 150 mm and 350 × 250 mm.

The refuse chute connecting the hoppers to the storage chamber is a straight vertical tube offset not less than 60 degrees at the foot. The walls are of solid brick or concrete a minimum of 100 mm thick with an internal diameter not less than 450 mm and lined with glazed ware or other impervious material.

Fan-coil unit in the ceiling

Exposed air-conditioning ductwork at the Hong Kong Arts Centre

垃圾收集工作

《建築物（垃圾及物料回收房及垃圾槽）規例》中清楚列明設施的標準與應用設計的需要。

在實用面積超過 1,320 平方米的設計中，必須包括垃圾收集站；而在實用面積超過 3,960 平方米非住宅用途的設計中，亦需考慮加入垃圾收集站。至於停車場、工業用途的建築物、學校、教堂等是可以申請豁免這項規定的。在較大型的計劃項目裏，還需要考慮到垃圾車行走的路線與需要。

垃圾站是一處讓垃圾車到達接收垃圾的地方，必須建有最少一道的外牆；在垃圾房中不可開窗及房間所有寬度要超過 1.5 米，房間淨高在 2 米以上。通常垃圾站的建築物料為磚塊、混凝土磚或混凝土。垃圾房內的牆身必須鋪砌容易清潔的光面瓦瓷；至於天花方面，以白英泥批盪抹灰；地台則鋪缸磚，地面要有足夠的斜度，可疏通清潔的污水。垃圾房內要裝置水龍頭。以密封鋼片製成的門的闊度最少有 12 米，及最高最少有 1.8 米。

設置於每層樓的垃圾收集槽要經常保持通風，槽內壁不可有任何突出的地方而使垃圾積聚。垃圾槽漏斗用鋼製造，最少要為 3 毫米厚的不銹鋼或鉛水生鋼片，橫開的門口須在 250×150 毫米至 350×250 毫米的標準。

垃圾槽管道必須畢直，到達垃圾站底時轉曲的弧度不能少於 60 度。垃圾槽的牆須至少 100 毫米厚，內部直徑要在 450 毫米以上，並塗上光滑或非滲透性物質作為表面處理。

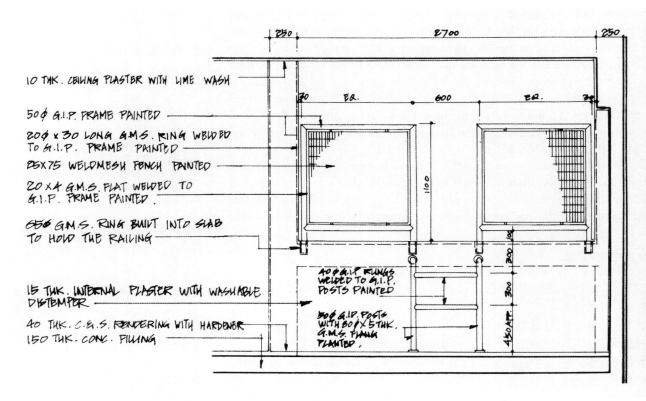

Elevation of cat ladder and railing in the lift machine room

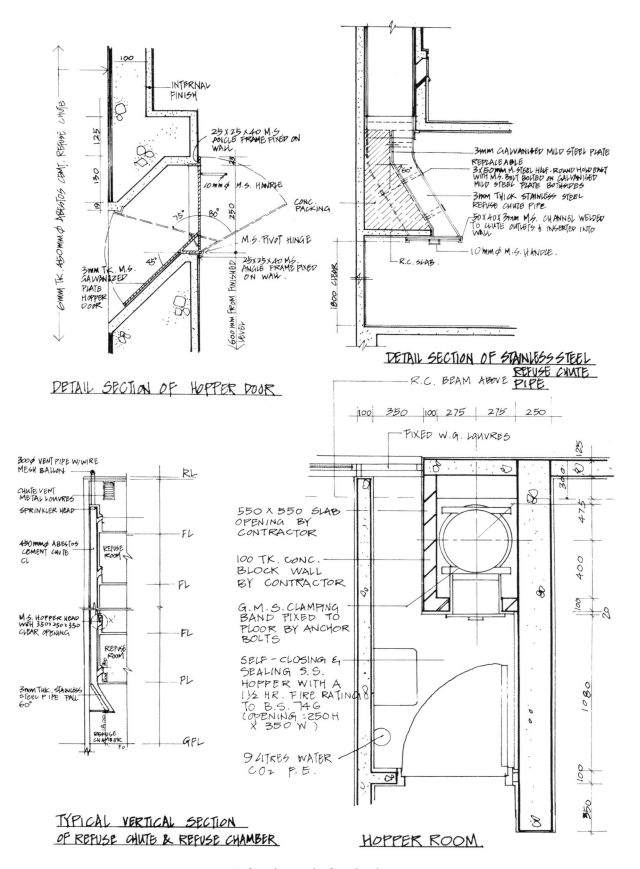

Refuse chute and refuse chamber

2.16
External Work and Landscape Work
外圍工作與園境計劃

Landscaping at the Simon Fraser University, Burnaby, BC

Planting of trees along the pavement with tree grilles

Landscape work

Landscaping is often the final touch of construction work. For large sites, planting, watering, and fertilization proposals often have to be considered by the Lands Development for approval.

The subsoil used is fine-grained decomposed granite free from impurities, the general depth of which is at least 300 mm. The topsoil is a fertile layer of free-draining material of a sandy loam character made of 3:1 decomposed granite and peat moss free from grass or weed, 150 mm thick. Planting season is from April to August for best growth of plants. Preplanting fertilizer of a slow-release compound comprising nitrogen, phosphorus, and potassium (10:15:10) can be added during cultivation of soil before planting. Plants are well watered before they are removed from containers for planting. Trees may need tripod staking for support. After planting, the plants are watered thoroughly. Fibrous organic mulch 75 mm thick can then be applied.

Depending on age and size, trees are classified as seedlings, whip trees, light standard trees, standard trees, and heavy standard trees. Seedlings are one to two years old and 150 to 600 mm tall. Heavy standard trees are over four years old and over 3,500 mm high. Shrubs are transplanted seedlings or rooted cuttings with a bushy appearance. Ground cover plants are perennials with low, close growth covering the ground surface. Climbers are plants maintaining an upward growth on walls, fences, etc. Turfing is green grass free from weeds, pests, or disease. General species are carpet grass, Bermuda grass, centipede grass, and field grass.

Hydroseeding is applied on large areas or on slopes. Seed mixes contain Bermuda grass, bahid grass, rye grass, rhodes grass, weeping love grass, centipede grass, buffel grass, or a combination of these.

Seeding is usually done by spraying in damp, overcast conditions. Sprays are blended with mulch, fertilizer, and a soil-binding agent. The sprayed areas are protected from strong sunlight or heavy rain by muslin or nylon net until seeds show good germination.

園境工程

園境工作一般視為建築工程中最後的項目。在較大型的地盤裏，種植、澆灌及施肥的安排，通常都要經地政總署的批劾。

底層土壤是較微細的、不含雜質的化解麻石碎。普通的深度為最少 300 毫米。表面的泥土是營養面，含有沙底之類的滲透性物質，由 150 毫米 3：1 的化解麻石與不含細菌的菌草碎合成。最佳的種植季節在每年的四月至八月間。在未正式種植時，可在泥土面預先噴上營養合成物，其

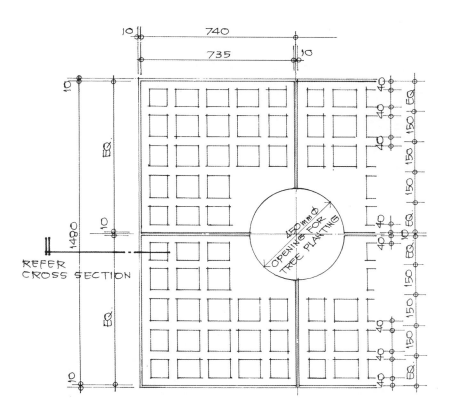

PLAN

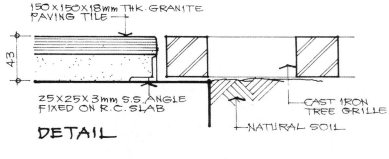

DETAIL

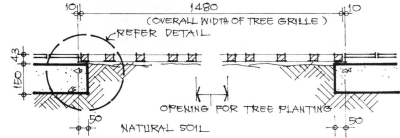

CROSS SECTION

DETAILS OF TREE GRILLE

Details of tree grilles

Podium garden at the Hong Kong Convention and Exhibition Centre

Triangular motifs at the Bank of China

Plane of water at the Bank of China

Chinese-style landscape

中包含氮、磷、鉀（10：15：10）。植物離開苗箱時，必須仔細澆水。樹木要以三邊的支援架模穩固其主幹。種好的植物，必須重新澆水一次，可在上面蓋上75毫米的有機纖維模。

樹木類別可按其年齡及體積分為：樹苗、狹木、輕盈標準的樹木、標準樹、重型的樹木頭。樹苗生長了一至兩年，高150至160毫米不等。重型樹木生長了四年以上，高度超過3,500毫米。灌木由苗種或切根法培植而成，具有枝葉較燦爛的結構外觀。地被植物是矮小類的地面生長植物，攀緣植物可以附著牆身與圍牆類植物攀爬而生。草地屬不含雜草、細菌等的青草。

噴草用於較大的地方與山坡位置，工作一般在較潮濕的天氣進行。噴草混合許多不同的護根、營養料與土壤的結合料等。噴草範圍須小心保護，避免受陽光與雨水破壞，完成噴面工作後，要用尼龍布覆蓋好，直至小草發芽生長。

Roads

Large sites are frequently required to construct their own private streets (a footpath on each side) or access road (a footpath on at least one side). The planning and construction of private streets and access roads is governed by the Building (Private Streets and Access Roads) Regulations. The width of the street depends on the type of use of the area and is determined by the Building Authority. The design of the kerb radius, road junctions, gradients, horizontal curves, and vertical curves is governed by the regulations.

The carriageway is constructed of a camber of 1 in 40. The surfacing of the carriageway is generally either of the following:

1. Concrete not less than Grade III and not less than 150 mm thick laid on at least 75 mm hardcore. A 75 mm thick BRC (British Retail Consortium) mesh-reinforced wearing slab can be added on top for prolonged use.
2. Bitumen macadam not less than 75 mm thick laid on hardcore not less than 200 mm thick with a finishing coat of fine bitumen macadam not less than 25 mm thick.

Kerbstone is constructed of granite (from China) or precast Grade III concrete 150 mm wide and less than 750 mm long. The depth of kerbstone is 300 mm and is partly embedded in the ground so that the top of the kerb is between 75 mm and 175 mm above the level of the adjacent channel.

The footpath is generally surfaced with concrete not less than Grade III, at least 50 mm thick, and covered with granolithic paving not less than 12.5 mm thick. Precast concrete tile paving on a compacted grade is also acceptable. Paving tiles such as artificial granite tiles on concrete slab is another alternative. The footpath is constructed with a cross-fall towards the kerb of 1 in 40.

The road is provided with surface channels constructed of Grade III concrete not less than 150 mm thick or 300 mm wide. In cross-section, the channel is laid to a fall of 1 in 30 towards the kerb and in longitudinal sections a fall not less than 1 in 100 although 1 in 250 may be permitted in special cases.

Road markings and traffic signs also form part of the road construction system.

Gateway at the Hong Kong Park

Small landscape work at the ground level of an apartment

Landscape lighting

Intimate bridge element

Swimming pool at the Greenwood Garden

Fountain feature

Flagpole and palm trees at the HSBC headquarters

Play equipment

Space frame gateway to the clubhouse

Landscape wall feature

Covered walkway at the Hong Kong Convention and Exhibition Centre

Artifical stone at the Hong Kong Convention and Exhibition Centre

Fence wall incorporating planters at the Greenwood Garden, Sha Tin

Construction of road and pavement finished with artifical granite

Road markings and pavement finished with artifical granite

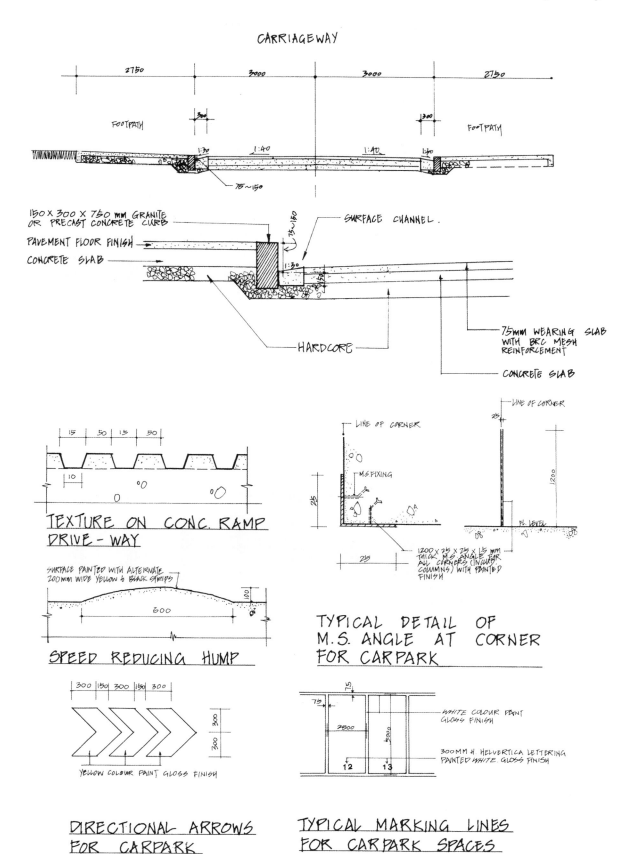

Car park and road markings

道　路

在大型地盤的地契裏，通常要求建築私人的「街」（兩旁有行人路）或「路」（只須其中一邊設置行人路）。在計劃與建築時，須參照《建築物（私家街道及通路）規例》。街道的闊度取決於發展的用途後，再經建築事務監督批准。至於行人路面的直徑、路的分叉口、斜度、橫向弧度、轉角弧度等，都有法例依據。

Fence wall

A fence wall not less than 1.8 m high is required by the Building Authority to be constructed around the site of a building adjacent to a street or lane.

The footing of such walls is usually an L-shaped cantilever structure. If construction is of brickwork or blockwork, the thickness is 100 mm for under than 1.8 m high and 225 mm for over 1.8 m high. A length of over 2 m is provided with 225 mm² buttresses or piers at not more than 2 m centres.

Other construction can be of concrete, steel wire, glass block, chain link, and so on.

圍　牆

根據政府建築條例，圍牆的高度必須要在 1.8 米以上，且圍繞地段的範圍界線內。

圍牆的一般地基為簡單的 L 型懸臂式結構。由磚或混凝土磚建築，如果牆身高度為 1.8 米以內時，牆的厚度為 100 毫米；而牆身高度在 2 米以上時，須以 225 毫米厚度的方形柱墩，每 2 米的距離內加在牆身。

其他的建築用料包括混凝土、鋼線、玻璃磚和鋼鏈等。

Chapter 3
Case Studies

實例研究

Wong Wah Sang 黃華生
Chan Wing Yan, Alice 陳詠欣

3.1

The Forum, Hong Kong, China: Curtain Wall Design
中國香港富臨閣：玻璃幕牆設計

The forum viewing from the podium

The Forum at Exchange Square was originally a three-storey retail building of concrete, which became obsolete upon completion of the new IFC shopping mall at the adjacent site. The client requested a new iconic Grade A office building surrounded by an improved outdoor public plaza connecting the footbridge and the shopping mall beyond.

This redevelopment offers a great opportunity for the architect to create a new landmark in Central. One key design consideration was the quality of public space created and the massing in response to pedestrian flow around the plaza. The final concept expressed the building as a precious urban 'gem'. The challenges of this project are of several aspects: the site constraint, limits of the existing base to carry the load, and statutory requirements.

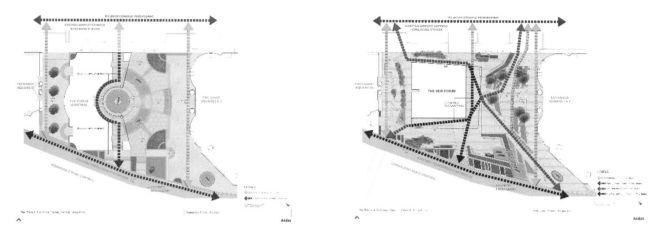

Diagram comparing the circulation pattern between the existing site force and the proposed new plan

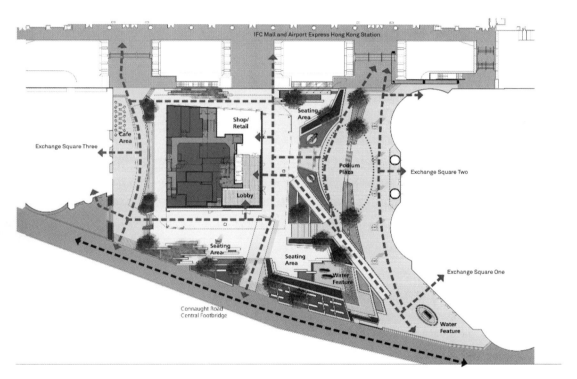

The master layout plan for the Forum

此項目為重建交易廣場發展計劃的其中一幢大樓,原大樓是一座三層高的混凝土結構商場。在旁有新建成的大型商場,故用處比以往大大減低。項目旨在將其重建為一座五層高、更為簡潔的甲級商廈,而且更能有效地連接廣場及其他周邊建築。

富臨閣重建項目提供了最好的機會,去打造一個位於中環優越地段的新地標。建築師提出其中一個主要的考慮,就是公共空間的質素及建築的形態要如何改善附近的人流交通。最終的設計構思是將大樓視作一顆珍貴的城市寶石,項目的形態創造了一個開放的空間,切合最初的設計原則。這個項目面對幾個挑戰:場地限制、原本地基荷載限制和建築物條例要求。

Site constraints: On structural and construction aspects

The building was planned to sit on an existing concrete podium deck, which is directly above an operating Public Transport Interchange (PTI). To avoid affecting the operation of the PTI, no alteration of the existing structure was allowed. The loading capacity of the existing podium structure was a concern to the architect and engineer to cater for the new building on top.

The main structural support points and the building core are strategically positioned on the original podium main girder. A lightweight steel diagrid structure was developed to distribute the loading uniformly onto the original podium structure. The nodes of the diagrid were set out to match the building floor levels to allow beams spanning between the diagrid nodes to allow for a structurally efficient design.

To avoid affecting the operation of the PTI, some of the major construction process or electrical and mechanical connection can only be planned at night, four to five hours per day. A packed programme was prepared. The construction period for the whole building was planned to be within 16 months.

場地限制：結構和建造方面

該建築規劃在現有的混凝土平台結構上，平台的底部有一個公共運輸交匯處。為了避免影響對交匯處的日常運作，現有的平台結構要保持不變，原結構的支承荷載能力是建築師和工程師需要考慮的一個關鍵問題。

最終採用了一個簡單的輕鋼斜肋結構，並將新建築的荷載平均地分佈在原有平台結構上。另一方面，為了避免影響公共運輸交匯處的運作，一些主要的工程進度及機電連接只能在晚間進行，每天只有四至五個小時。承包商和場地管理商制定了項目流程，項目建設計劃在 16 個月內完成。

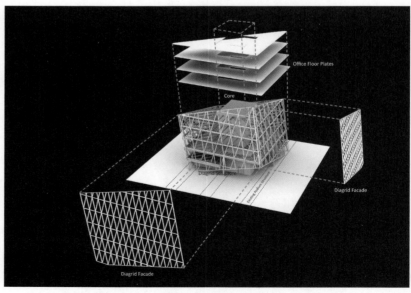

An exposed diagram showing the structural system of the Forum

Diagrid system: Importance of temporary support

Differing from an ordinary column and beam structure, the diagrid system cannot stabilize itself until the steel framework is completed and therefore requires a temporary support frame. In order to release the temporary support well, a specially designed 'sand jack' is used at the interface between the temporary support and the podium deck. A sand jack is basically a steel tray filled with densely packed sand, which bears the loading from the temporary support. After the grid is completed, the steel tray is dismantled and the sand is removed bit by bit, to lower the temporary structures before they are detached last. The permanent steel structures are designed to have a certain precamber value. Upon completion of the building, the building reaches its design dead load, and the settlement of the structure brings the building precisely back to its design level.

斜肋結構架：臨時支撐的重要性

與傳統的柱梁結構不同，斜肋結構在鋼結構未完成時是不能達到自我穩固的，所以需要臨時支撐架來穩固。為了更容易的拆掉臨時支撐，特別設計的「砂箱」用於平台和臨時支撐的接口。「砂箱」是由一個裝滿砂的鋼鐵托盤組成，用來承受臨時支撐的負荷。當斜肋結構完成後，鋼鐵托盤會被拆卸。砂則會一點一點的被轉移來降低臨時結構，最終它們會被清除。最終使用的鋼結構有一定的預拱度，當建築完成時，建築達到預計的淨負荷時，就會返回設計時的特定水平。

The construction of a diagrid structural frame

Façade design: Modular design concept

The form of the building is a gem in the landscaped plaza of Exchange Square. The tilted building form was designed to be supported by a diagrid structure for stability and running in line and integrated with the façade mullion. A unitized curtain wall system was used. Each curtain wall module is made up of two triangles which form a parallelogram shape. The modular dimension is 2.4 m × 4.5 m, which is in line with the floor-to-floor height. Much consideration was given to the size of the module to match the office modular planning grid and allowed the node of the diagrid to coincide with the floor plan.

外立面設計：模組設計概念

項目的概念是一個在交易廣場平台花園中的菱形寶石。當設計外立面時，菱形網格結構作為一個框架來承托混凝土樓板，外立面則採用單元式幕牆。每個幕牆模組是由兩個三角形組成，形成一個平行四邊形，以配合斜肋鋼結構。該模組的大小經過詳細的考慮，寬度為 2.4 米，高度則等於一個 4.5 米的樓層。

Curtain wall glass panel

The selection process of the glass for the external façade is important, as it will greatly affect the appearance of the 'gem'. The glass panel of the curtain wall is an IGU consisting of two 12 mm glass panels with a 12 mm air gap. The outermost glass is a heat-strengthened glass panel, and the inner one is a tempered-glass panel. The use of heat-strengthened glass has the advantage of better surface flatness. Also, considering the OTTV of the building, a low-e coating is used to give a lower shading coefficient to the glass. For colour, transparency, and reflectivity, a mock-up of different glass types is hoisted into the air and tilted at the exact angle of the building façade to simulate the as-built lighting and reflections for the client and the architect to select.

Details showing the façade design

The sealant joint design

Mock-up for the selection of glass

For the façade, the tilting angle of the massing is aligned with the modules so that the transom and mullion can be touching each other in perfect shape.

As the architect wanted to expose the diagonal members, the transoms were concealed and the diagonal members were highlighted with aluminium capping. The gaps between the panels were not filled with gasket but with aluminium expressed. Movement was allowed in the detail design.

玻璃幕牆

玻璃的選擇非常重要，因為它在很大程度上影響到「寶石」的外表。幕牆的玻璃面板是由兩片 12 毫米厚的玻璃與 12 毫米空氣間隙組成的中空玻璃。最外層的玻璃是半鋼化玻璃，內層是全鋼化玻璃。使用半鋼化玻璃的好處是可達至更平坦的表面。此外，考慮到建築物的總熱傳送值，使用了低幅射塗層以減低玻璃的遮陽係數。體量的傾斜角實際上是與模組對齊，使得橫框和豎框以完美形狀相互接觸。

由於建築師設想構件表現出斜線，故橫樑被掩蓋，而斜構件則在外面加上鋁覆蓋。為了強調線條，面板之間隙沒有裝置墊片而凸顯了斜線鋁組件。細部設計則留意建築物的微移動。

Construction of a composite metal deck

Construction of composite deck slab

With the use of a structural steel frame, a composite steel deck was used for the floor slab system. This is a system of structural decking with profiled steel sheeting, which performs as part of the structure and a permanent left-in formwork for the slab. After installation of the deck, the reinforcement was set up and concreting followed.

鋼板加混凝土組合樓板的建造

配合鋼結構框架，樓板結構採用了鋼板加混凝土組合樓板。預製壓型鋼板成為永久性樓板的底部，然後進行鋼筋加固和混凝土澆灌。

Coordination with consultants and contractors with BIM

The construction period for the project was short and because of its steel structure, a Building Information Model (BIM) was introduced to make sure that all the architectural design, structures, and E&M services were properly coordinated. Any mistake or clashing of the service destroying the cleanliness of the building is avoided. Some of the piping, like water pipes for sprinklers, was designed to pass through the steel beams to save space, also well noted before prefabrication. All piping and routes were drawn in the 3-D model so that consultants could use the model to fine-tune the final product. With the help of the latest software, the contractor could also plan all the temporary work and safety work surrounding the site and easily show it to the client.

協調顧問和承包商的建築資訊模型

由於項目為鋼結構，加上項目建設期很短，建築資訊模型（BIM）被引入應用，以確保所有的建築設計、結構和機電設備的配合。不同組件的任何錯誤或衝突，都會破壞建築設計。譬如消防灑水器水管的管道穿越了鋼樑以節省空間，所有的管道和途徑都被繪製在三維模型上，使顧問能根據模型進行最終的微調。利用三維模型，承包商也可以用來計劃所有臨時工程和工地安全的包圍工程，並很準確地以圖像與客戶溝通。

Client:	Hongkong Land
Architect:	Aedas
Structural Engineer:	Ove Arup & Partners Hong Kong
Mechanical and Electrical Engineer:	J. Roger Preston Limited
Landscape Consultant:	Aedas Landscape
Management Contractor:	Gammon Construction Limited
Year of Completion:	2014

3.2
CIC Zero Carbon Building: Eco-Building Design and Technologies
零碳天地：環保建築及技術

An overview of the Zero Carbon Building

The CIC Zero Carbon Building is the first zero carbon building located in the subtropical and high-density context of Hong Kong. Beyond the common definition of a zero carbon building which offsets operating energy consumption by on-site renewable energy generation on an annual basis, it exports surplus energy to offset the embodied energy of the construction process and major structural materials. The architectural design emphasizes energy consumption reduction by state-of-the-art passive design and active systems. It also generates on-site renewable energy more than operation needs, from photovoltaic panels and a biodiesel tri-generation system.

To showcase the eco-building design and technologies and to raise public awareness of sustainable living in Hong Kong, real-time control and monitoring was done by a comprehensive building management system with user-friendly graphical interface, called Building Environmental Performance Assessment Dashboard (BEPAD) with over 2,800 sensing points.

The site also successfully created the first urban native woodland in Hong Kong, with over 300 trees of over 40 different native species, to enhance biodiversity, providing food and shelter for native wildlife in the city.

零碳天地是香港這個亞熱帶高密度城市的第一座零碳建築。除了零碳的基本定義外，這座建築能把每年的可再生能源發電用來抵消能量損耗，並且將過剩的能源用來抵消施工過程和主要結構材料的蘊含能源。同時，這座建築主要由最先進的被動式設計和主動系統來減少能源損耗，還有經光伏板和生物柴油聯產系統製造超過建築運行所需要的再生能源。

為了展現環保建築設計和技術，以及提升香港公眾的可持續生活意識，整個項目使用方便用戶的圖形界面綜合建築管理系統：在各個角落提供超過 2,800 個傳感點，並連繫到建築環境性能評估儀表板（BEPAD），從而實施實時的監測和控制。建築場地還成功創造了香港首個城市原生林地，擁有超過三百棵及四十多種本土樹種，來提供生物的多樣性，以及提供食物和庇護予城市裏的野生動物及雀鳥。

Structural design

A shallow foundation was used because of its low cost and the light structure of a two-storey high building. Balanced cut and fill was designed by the structural engineer to minimize import or export of soil to reduce embodied energy during construction. Pulverized fuel ash (PFA) was added to the in-situ reinforced concrete structure. Also, for minor reinforced concrete, 20% of recycled aggregates were used for landscape features, planter walls, and minor footings.

結構設計

大樓為兩層建築，使用了低成本且輕巧的淺基礎結構。結構工程師為了減少地盤平整的土壤進出，盡量平衡土地的切割及填充。同時，加入了粉煤灰於鋼筋混凝土中。其中一些比較次要的鋼筋混凝土結構，有 20% 為再用石材，應用在園境設施、花槽和次要的地基。

A typical section showing a tall window for sunlight

Building form consideration

The design of the building form and orientation were considered to enhance the energy capture on the roof, cross-ventilation, a stack effect for natural ventilation, and harnessing of day lighting. Site information such as the sun path and prevailing wind directions was collected and acted as a basis for the building design. The built form, with a leeward roof, works with the prevailing wind to create a large negative pressure and helps to draw natural ventilation through the building. Niches and recesses of the built form facing south-east are designed to increase the passive zone with good ventilation and infiltration of day lighting.

No air-conditioning installation was required for the permeability of the high building and the adaptable space of the main entrance lobby, which has comfortable shade and fully open sides facing the prevailing wind.

建築體量設計

為了提高屋頂的能量利用率、空氣對流、自然通風的煙囪效應和採光收穫，建築師特別考慮到建築的形態和方向。地形的基本資料，例如太陽軌跡、盛行風向等，都被考慮作為設計的基礎因素。建築的外型，利用了順風的斜屋頂與盛行風能，形成一個巨大的負壓力，用來幫助建築物自然通風。建築東南面的凹入設計，是用來增強被動空間的對流空氣和陽光滲透。

主入口大堂為一個高透氣性和高活動性的空間，即使沒有空調的安裝，只有一些遮陽裝置及一個完全開揚的空間，已能提供一個舒適的入口給遊客及使用者。

Window-to-wall ratio and glass panel

A relatively higher glazing ratio (approximately 65%) for views as a transparent interface with the neighbourhood was proposed for the street façade facing Sheung Yuet Road. The visual transparency of the façade is enhanced by tilting the glass wall. The south-east façade with a high window-to-wall ratio with overhang offers shade from the summer sun while allowing good views of the landscaped area. The west and east façades are relatively opaque, and the roof

The entrance of the visitors' centre

is shaded by photovoltaic (PV) panels and a green roof. The OTTV is calculated to be 11W per m², which is about 80% better than the current statutory requirement.

窗牆比例和玻璃牆板

面向常悦道的外立面有相對高的窗牆比例（大約 65%）。建築師利用傾斜的玻璃牆，增強外立面的透明感。東南面有較高窗牆比例，以提供更多的園林景觀給遊客，同時以懸吊的遮陽頂來減少陽光的直接照射。西部和東部外立面相對不透明，屋頂則由太陽能光伏板和綠化遮蓋。總傳熱值（OTTV）為 $11W/m^2$，比目前法定最低要求超出近 80%。

Architectural and construction details

There are several details implemented by the building, contributing to sustainability for the project.

Inverted beams for roof and photovoltaic panel

The inverted beam at the roof suggests a flat soffit for the building, which has a white surface facing the north light so as to diffuse daylight into the interior. The beams on the roof level also reduce the extent and material use of the metal supporting frame for the PV panels. Different kinds of PV panel were installed in different locations. Polycrystalline PV is used on the main roof. The modular unit assembly system on the roof is designed to cater for any mounting situation or configuration to provide flexibility for the choice of different PV modules in the future. A building integrated panel was used for the glass roof of the viewing platform. The cylindrical copper indium gallium selenide (CIGS) PV collects direct, diffused, and reflected sunlight from 360 degrees with high rooftop coverage and without the need for mounting hardware. The system was estimated to produce about 85 MWh renewable energy per year.

Active skylight

The external fin blades at the skylight are motorized and actuator-controlled to vary the amount of natural sunlight according to daylight data collected from the sensors and with respect to the sun path of Hong Kong and the sky conditions of the site. Control panels are connected to the Building Management System (BMS) for monitoring of the damper and power supply status.

Electrical operable window system

The electrical operable window system comprises window actuators, window controller, window status contacts or sensors, and other necessary wiring and accessories. The control of actuators is under the same BMS with a manual override option. The convenience and intelligent control of windows enhances the applicability of natural ventilation and improves the control of windows in less accessible locations.

The following details also add value to the sustainability of the project:

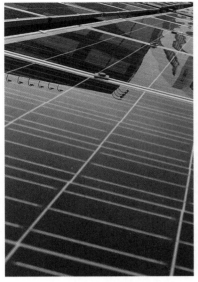

Multi-crystalline on the inclined roof

BIPV at the viewing platform

Cylindrical CIGS at the Eco-café

Active skylight on the inclined roof

1. Wind catcher. Regulates fresh air intake by motorized dampers and sensors for internal temperature, CO_2, wind movement, and humidity under the BMS or manual control.
2. Light pipe. Use of clear polycarbonate roof dome with UV inhibitors and aluminium tube laminated with a silverized Polyethylene terephthalate (PET) film to reflect collected light to the interior. Light levels controlled by motorized dampers through light sensors are under the BMS or manual control.
3. Earth cooling tube. Underground ductwork for fresh air intake and pre-cooling with fan-assisted ventilation. As the temperatures of the underground earth mass are relatively stable and lower than the average air temperatures in summer, the pre-cooling can be done by the cooler thermal mass.
4. High-volume-low-speed fan. The fan moves a large amount of air at a slow speed. The patented airfoil and winglet are designed for efficient airflow by eliminating vortex formation at airfoil tips. An acceptable temperature for inhabitants is raised by 2°C for an airflow of 0.5 m^2 from the fans. The overall energy saved would be 14% in the building.
5. Biodiesel tri-generation. The device makes use of the biofuel from waste cooking oil for cooling, heating, and power generation. The building requires approximately 50 tonnes of biofuel per year, approximately 2% of local production capacity in 2009. As the biofuel is made from waste cooking oil produced locally in Hong Kong, the emission factor is comparatively lower than the combustion of fossil fuel and thus avoids the generation of methane gas at landfills. It is estimated to produce about 145 MWh of renewable energy per year.

The Zero Carbon Building involved quite a lot of new technology in construction. Most of the technology was the first application in Hong Kong. It is the first zero carbon building addressing specific environmental challenges in the local building industry and marks a new period in sustainability technology.

建築和構造細節

以下為項目採用的若干細節，有助於可持續性發展。

屋頂倒樑和太陽能光伏板

項目的屋頂使用了倒樑,於是最頂層的天花為一塊平坦的白色表面,這個白色表面能讓北面的光反射並進入室內。屋頂上高起的樑同時成為了支撐光伏板的金屬支撐架的一部份。不同種類的光伏板分別裝在不同的位置,多晶光伏板用於主屋頂。屋頂的模塊組裝系統適用於各種組裝配置,使未來可以選擇多種不同的光伏板,提供某程度上的靈活性。建築整合太陽能板則用於觀景台的玻璃屋頂。圓柱形的 CIGS 光伏從 360 度收集直射、漫射及反射的陽光,更不需要吊掛的裝置器具,該系統估計每年能產生約 85 MWh 的再生能源。

天窗外部的葉片可根據感應器收集的陽光數據,按香港的太陽軌跡以及場地的天空情況,自動控制不同的陽光量進入室內。控制板則連接著建築管理系統(BMS),用來管理所有風閘和電源供應的狀態。

電動開窗系統

電動開窗系統包含窗口制動器、窗口控制器、窗口狀態接觸或感應器和其他必要的佈線及附件。制動器的控制是由建築管理系統控制,當中也可通過手動操作。窗口控制的便捷性及智能性,大大增強了自然通風的應用性,而且提高了難接觸地方之窗口的控制。

以下細節同樣增強了該項目的可持續性:

(1) 捕風器:以機械風閘調節新鮮空氣的攝入來控制室內溫度、二氧化碳、風向及濕度,全部都經建築管理系統或手動控制。
(2) 導光管:利用透明的聚碳酸酯圓頂及鍍銀質熱塑性聚酯(PET)的膜鋁來收集陽光進入室內。光線強弱由機械閘通過光線感應器,由建築管理系統或手動控制。
(3) 地底預冷管:新鮮空氣經由地下管道系統進入,並以預冷風扇輔助流通。由於地質的溫度相對穩定,且溫度低於夏日的平均空氣溫度,所以能達到預冷的效果。高量低速風扇:風扇以慢速移動大量的空氣。這是高效的空氣流動專利翼型及小翼設計,消除在翼尖形成的漩渦。在 0.5 m/s 的空氣流速下,溫度提升可接受的兩度。整個建築的能耗節約達 14%。
(4) 生物柴油三代裝置:該裝置使用煮食廢油的生物燃料,用來冷卻、加熱和發電。建築工程每年消耗約 50 噸的生物燃料,在 2009 年大約為本地產能的 2%。由於生物燃料由香港本地的煮食廢油煉製,排放係數相較低於化石燃料的燃燒,且可避免填埋的沼氣生成。據估計每年能產生約 145 MWh 可再生能源。

零碳天地涉及相當多建造方面的新技術,而且大多是在香港首次應用;同時,它也是首個挑戰本地建造業的零碳建築,翻開了可持續發展技術的新篇章。

Client:	Construction Industry Council
Architect:	Ronald Lu & Partners (HK) Limited
Civil and Structural/Mechanical and Electrical/Sustainability Engineer:	Ove Arup & Partners Hong Kong Limited
Management Contractor:	Gammon Construction Limited
Year of Completion:	2012

3.3

Domain: Redevelopment of Yau Tong Estate Phase 4, a Sustainable Commercial Building

「大本型」：油塘邨第四期重建發展——推動可持續發展的商業建築

The major façade of the Domain Mall

Yau Tong Estate Phase 4 is the latest phase of a comprehensive redevelopment of Yau Tong Estate. The Housing Authority aimed to create a vibrant retail space through a design which responds to the site, the environment, and the local community. Supported by the environmental consultant, the retail centre has introduced a number of special environmental features contributing to sustainable development. They are as follows:

1. Conserving energy through hybrid ventilation system
2. Enhancing environmental comfort through roof greening, light tube, and solar control
3. Conserving water through the use of reclaimed water
4. Providing a clean environment for food-waste collection by installing a food-waste pulping system
5. Achieving Platinum Rating under BEAM Assessment

From construction, material selection to environmental systems and features introduced, the development of the Domain Mall takes a leading role in fostering a more sustainable community in architecture.

油塘邨第四期是油塘邨最新一期的綜合重建項目。房屋委員會的目標，是建設一個場地環境適合本地社區的活力及商業空間。環保顧問為商業中心引入了幾個特別的環保項目，促進以下幾方面的可持續發展：

(1) 混合通風系統節能
(2) 屋頂綠化、自然光管和陽光控制，以提高環境舒適度
(3) 使用再生水節約水資源；
(4) 安裝廚餘收集系統，為剩餘食物提供衛生的環境；
(5) 達到建築環保評估協會授予的白金級別。

利用環保選材、建造和各種環保特色，大本型商場的發展領導了可持續社區建築發展的基準。

Sustainability in structural and construction aspects

The building was planned to sit on the Mass Transit Railway (MTR) of Yau Tong station. The challenge to the architect and the engineer was to optimize the foundation design, making use of the as-built foundation as much as they could without adding too many excess structures. The BIM was introduced to sort out the integration of the old and the new foundation structure. Minimal resources were introduced with no abortive design; the construction was also planning to achieve BEAM 04 Platinum sustainability credits.

結構和建造上的可持續性

這個建築項目建在油塘港鐵站上蓋，對於建築師和工程師的挑戰是，如何在現有的結構基礎上，不用增加過多的結構而能改善地基系統。此項目利用建築資訊模型（BIM）整合新舊結構，用最少的資源得到最優化的設計。同時，建造時也計劃角逐建築環保評估協會可持續發展獎項的白金級別。

Sustainability in architectural design aspects

The building form, orientation, and layout were designed to optimize cross-ventilation and harnessing of day lighting. Micro-climate studies were carried out in the early design stage. Computational fluid dynamics (CFD) wind analysis was carried out to identify thermal comfort zones for various outdoor activities. Maximized wall openings were designed for natural ventilation in the carpark and PTI, so that operation of the mechanical ventilation system can be optimized. Also, the public open space, podium, and greening of the roof were maximized to address the heat island effect.

The use of recycled materials in finishing is also a credit to sustainability. Concrete pavers with recycled glass aggregates were used in the PTI; recycled timber was used in all outdoor boardwalks, decking, and grilles; recycled medium-density fibreboard (MDF) was used for interior spandrel panels in the atrium.

建築設計上的可持續性

建築的形態、方向及佈局設計，都是為了達到最好的對流通風和日光能源。在設計的早期就展開微氣候的研究。計算流體動力學（CFD）風向分析，用以辨別不同室外活動的氣候舒適區。最大的外牆開口給停車場和公共運輸交匯處（PTI）提供了自然通風，因此室內的機械通風可以減少。此外，項目盡量綠化公共開放空間、平台和屋頂，以減少熱島效應。

使用回收的材料同樣是為延續可持續發展的理念。於公共運輸交匯處，使用回收玻璃碎料於混凝土作鋪設行人路用；回收木材則用於室外木板路、戶外地板和柵欄；回收的中密度纖維板（MDF）用於中庭的樓層飾板。

Sustainability in building system aspects

The glass panel of the curtain wall is an IGU consisting of two panels of glass with an air gap in between. The advantage is isolation and reduction of heat absorption. Having an IGU in the curtain wall design means the use of air-conditioning can be minimized but at the same time retain the natural light harnessing, which in turn reduces the use of artificial lighting.

The curtain wall design is connected to the BMS for a hybrid ventilation system. Openable windows were installed for intake of cool ambient air and exhaust of hot air. When the outdoor temperature, humidity, and air velocity meet the preset criteria for a comfortable natural ambience, the automatic operation system will be activated to turn off the air-conditioning and open the windows. Natural ventilation is usually provided during spring and autumn. The resulting annual operation hours of natural ventilation are about 582, contributing to 13.3% of the total operation time of the shopping mall.

建築系統上的可持續性

幕牆的玻璃面板是由兩塊玻璃和中間空隙組成的 IGU（中空玻璃），有隔離和降低吸收熱能的優點。幕牆裝置中空玻璃，有助減少使用空調，同時保持吸收自然光，減少使用人工照明。幕牆的設計連接到混合通風的建築管理系統，可用開啟的窗戶攝入冷卻的鮮風和排除熱空氣。當室外溫度、濕度和空氣流速達到舒適的預定標準，自動作業系統將啟動去關閉空調和打開窗戶。自然通風通常是在溫和的春季和秋季提供。每年的自然通風時間約為 582 小時，為商場運營時間的 13.3%。

The use of an MDF panel for the spandrel design

Bird's-eye view of the Domain Mall

Water conservation

Condensate water collected from the air-conditioning system is used for irrigation of landscaped areas. The air-conditioning system per day can irrigate 8,700 m² of green area. The estimated cost savings for water consumption is around HK$19,000 per year.

節約用水

從空調（A/C）系統所收集的冷凝水會用於景觀區的灌溉。空調系統每天可以澆灌 8,700 平方米的綠化面積，每年節省用水的成本估計約為港幣 19,000 元。

Reuse of kitchen waste

Pulpers were installed in the refuse storage and materials recovery room from the first floor to the third floor of the shopping mall. Kitchen waste is collected regularly by the food-waste pulping system and delivered to a laboratory for manufacturing of spent mushroom compost (SMC). The SMC is then transported back to the PTI area of the shopping mall, growing in panels to improve the air quality in the PTI.

廚餘回收

碎漿機安裝在購物中心第 2 至 4 層的垃圾及物料回收房。廚餘垃圾由廚餘垃圾回收系統定期收集，並提供給實驗室製造蘑菇堆肥（SMC）。堆肥會運回商場公共運輸交匯處生長面板上，以改善該處的空氣質素。

Solar light tube system and solar panels

As a device to harness day lighting, the solar light tube system captures natural light for the toilets located below the roof garden. Also, solar panels are used to generate hot water for the baby care room in the shopping mall.

This project has set a very good example for commercial buildings to promote the concept of sustainability to the general public. As the shopping mall has a close relationship with the community, the architect and the interior designer put great effort into the public area to show environmental awareness in order to educate our next generation. And in return, the concept of sustainability makes this mall standout from ordinary shopping centres.

太陽光管系統和太陽能光伏板

作為一個利用日光的設備，太陽光管系統既為位於屋頂花園下方的廁所捕捉自然光，也為購物中心的育嬰室產生熱水。作為一個商場項目，能落實可持續的概念，能大大提升商場的層次。建築師及室內設計師也著重於公眾地方，展示及推廣環保意識給社區人士，讓他們更加留意可持續的概念，並以此教育下一代。

Client: Hong Kong Housing Authority
Architect: Hong Kong Housing Authority
Civil and Structural Engineer: Meinhart Ltd.
Management Contractor: China State Construction Engineering (Hong Kong) Ltd.
Interior Architect: Aedas Limited
Year of Completion: 2012

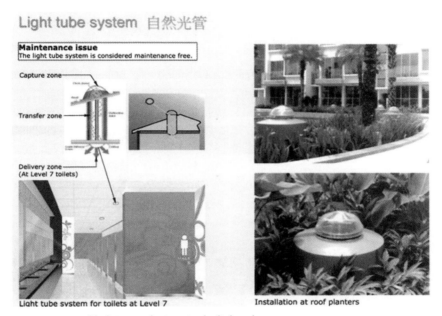

Bringing natural lighting to the interior by light tubes

3.4
Harmony 2, Tin Yiu Estate, Phase 3, Tin Shui Wai: Prefabrication Works
天水圍天耀邨和諧式的設計組合及預製模件

Tin Yiu Estate in construction

Harmony Blocks, as the name implies, symbolize community living where people live in harmony with their neighbours and the environment. Technically, this series of housing has been developed infrequent dialogue with the construction industry from the big concept down to small details. We shall see that, in this project, quality is the emphasis when the Harmony Blocks were completed in 1992.

理想的居所設計,是要配合人們的日常生活方式和喜好習慣。以更具體的層面看,是從設計中表現群體的活動空間,而且處處也能反映和諧的、自然的生活環境。

　　香港房屋委員會於 1992 年完成的天耀邨,是屬於和諧式的公屋設計,也正好反映上述所說的理想居所。

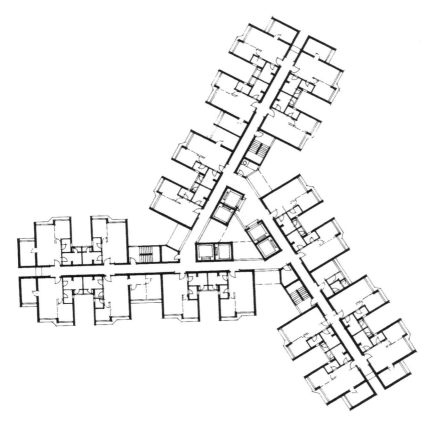

A typical floor plan

Modular design concept

A computer-aided design is the main feature in the design process of Harmony Blocks. A modular approach allows flexibility in the mix of flat size. One-bedroom, two-bedroom, or three-bedroom modular flats can fit into the planning of the blocks. Building components can also be standardized, allowing prefabrication and better quality control during production. This also reduces the labour requirement on site.

Harmony 2 is a 37-storey trident-shaped tower with 18 flats per floor. The trident-shaped plan affords less likelihood of residents looking into other flats and thus more privacy. The three identical wings allow rotational and repetitive use of formwork.

By staggering the central lift lobbies in three alternating floors, a six-storey high atrium is formed at the centre of the block, giving the neighbourhood community the space for a high-rise residential building.

Ancillary facilities are located on the ground floor. By grouping services installation in one wing, the remaining ground floor space can be used as the main entrance and for non-domestic amenities.

模組設計

電腦科技的發展，是推廣和諧式設計的主要副手。重複的單位和係數式的設計組合，令模組設計的使用者有很大的彈性和自由度。由小漸進，

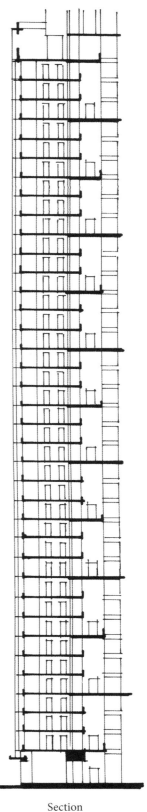

Section

From laying reinforcement to concreting of the floor slab arranged by rotation of the three wings

每個小單元的組合,配合建築裏的各項設備,成為一個獨立的模型。而模型本身的設計,當然亦要以最理想的標準來訂定,餘下的只不過是組合的問題。這樣的好處是:

(1) 提供高效率的生產線;
(2) 減低地盤工人的數量,同時提高物料的質素。

所以,從建築和經濟效益的角度來看,模組設計的概念將是建築業的新里程碑。

天耀邨的和諧式公屋,是一座三十七層高的 Y 型設計,每層共有十八個單位。Y 型設計的優點是盡量避免住戶間互相對望,使各單位保持獨立性。

設計的另一個特點是中央的運輸設計,每三層轉換電梯的位置,可以創造六層高的前庭(atrium),提供一個舒適的開放空間,較一般的高層住宅樓宇為罕見。

建築機電和各類別的設施,集中於底層三個翼的其中一個,而剩下來的,可作為公眾休憩地方,提供更多地面使用空間,也是天水圍公屋的最大特點。

Load-bearing walls

The basic structure comprises groups of shear walls at the wings and lift core walls at the centre to resist lateral wind load.

Large panel construction in the form of table and wall formwork is used for casting the floor and structural shear walls. This will save time in erecting, striking, and re-erecting the formwork, as the formwork is handled as one unit. However, the reinforcement is still worked by hand, cutting, bending, and shaping into the required size and form. The table form can produce a surface neater than can conventional plywood formwork and practically needs no applied finishes for levelling. The outside wall is further added with spatter dash before applying the mosaic tiles. The concreting cycle for one floor is nine days.

承重牆的形式

從設計的平面圖中,可以看到建築的結構特色:中央的電梯間和一組一組的承重牆,承受了風力和大廈本身的重力。

大型的承重牆設計,是為了配合預製模件的建築程序。傳統的紮鐵和釘板之後,再由澆灌混凝土到拆板,約九天的時間便可完成一層。

A precast façade

One of the Harmony 2 blocks was used as a trial project for precast façade. Between the shear wall structure is the external façade, which incorporates the concrete spandrel, walls, and windows. Having prefabrication on the ground, better quality is achieved.

A casting yard of 20 m × 60 m was set up on site. Steel formwork manoeuvred by a travelling crane was used instead of the conventional plywood forms. The façade formwork was turned face down onto the ground.

System formwork for the construction of the shear walls. A working platform is incorporated in the steel formwork.

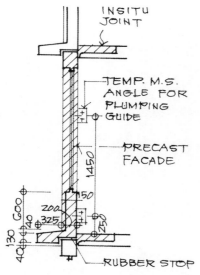

SECTION

The casting yard on site

The steel window frame together with the reinforcement bars were fixed in the formwork. Concrete was poured into the formwork and trowelled flat. Simultaneously, concrete cubes were made for strength testing and strength control.

After the concrete had set for seven days, the horizontal steel formwork was erected upright. After a further 14 days, the formwork could be demounted. Spatter dash was applied on the overhang portion where mosaic tiles were applied, and other portions were left smooth for spray painting.

The finished façade was then moved to another area for storage until a tower crane lifted the façade to be erected on Harmony Block. The façade was fixed into position by non-shrinkage grout. Part of the concrete slab adjacent to the façade was left unbuilt until the façade was fixed in position, after which this part was concreted to form a homogenous tie between horizontal and vertical planes.

The advantage is a fine-quality finished façade achieved by such a prefabrication method. The problem of water leakage with this type of construction is primarily dealt with by the overhanging toe of the whole prefab unit, which stops water penetration from the horizontal parts. The steel windows casted together with the spandrel and adjacent wall will theoretically reduce water seepage through poor grouting to the minimum. The only weak point lies in the vertical joint, which is secured by in-situ non-shrinkage grout.

Another interesting feature is the precast staircase, constructed in eight risers with recess to receive nosing tile together with a prefabricated balustrade panel. The half-landing can be casted together or separately as in-situ part.

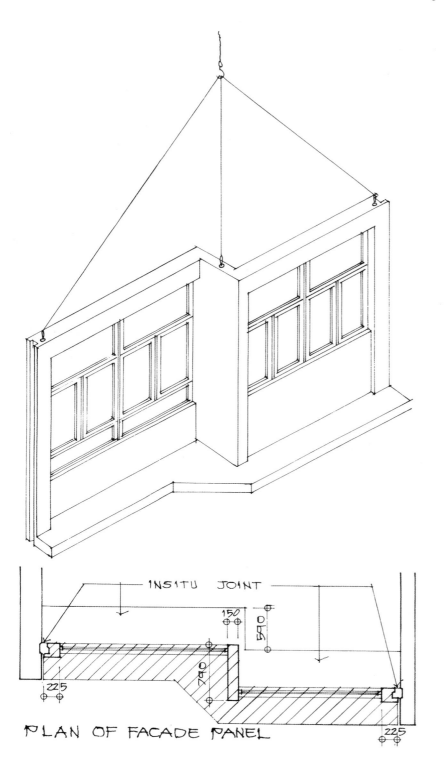

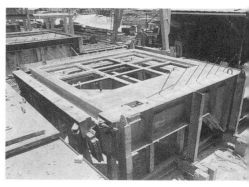

Prefabrication of the façade unit

Prefab façade

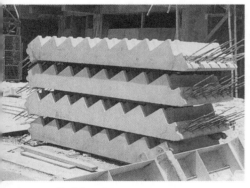

Prefab staircase

Prefab staircase assembled on site

預製的外牆

房委會在興建這個和諧式屋邨時,嘗試用一種新穎的建築方法。

在進行承重結構工程的同時,於地盤的另一個角落(約為 20×60 米的面積),進行鋼板模組的預製外牆作為一個單位的模式。由於在地面工作,工人可以較容易操作巨型和先進的機械鋼板,優點是較為耐用;同時,生產出來的外牆會更準確、較易控制其效果,因此,生產的質素也較為良好。況且,由於是大量的建造,從生產效益上看,其實一點也不昂貴。

完成後的獨立外牆,連窗、窗花及外牆的磚仔,被運送到每層的結構架上安裝。不過,此項建築方法的最大缺點,是介面之間的縫隙問題。除了用防水混凝土的連接劑外,每一層外牆模件的下部份,都設計了一個小遮擋來減低漏水的可能性。

還有一個有趣的預製模件就是八級的樓梯元件。從整體的概念看來,建築的程序與砌積木的玩兒也不遑多讓了。

Construction of windows and doors

Steel windows are standardized and arranged under various modules before installation in the casting yard. Standardized mild steel bars are also installed by screwing to angle cleat with countersunk self-tapping screws. Some of the transoms are designed to hold hangers and trays for installation of air-conditioning units.

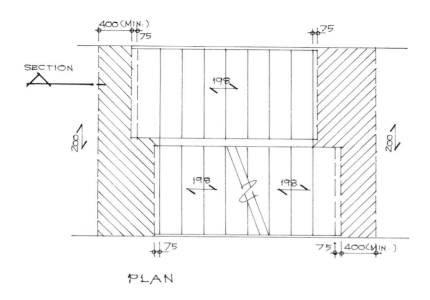

Prefabricated façade fixed on site with partial floor slab left for in-situ concreting

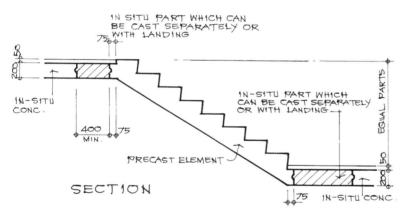

Steel windows with grilles installed

Plan and section for the precast staircase

Three types of standardized door are designed: 50 mm solid core flush door, 45 mm hollow core flush door, and 50 mm solid core flush door with a vision panel of wired glass. The doors are constructed of a high quality with lipping and luan veneer facing.

窗與門之製造

鋼窗在未運到預製外牆的工場時，已經在廠裏加工製造。而在地盤的工序只是把它們安放於模架上，然後澆灌混凝土。因此，品質和數量也可以控制。在某些窗門的花架上還預備了外架，讓入住後的家庭可以自行安裝空調設備。

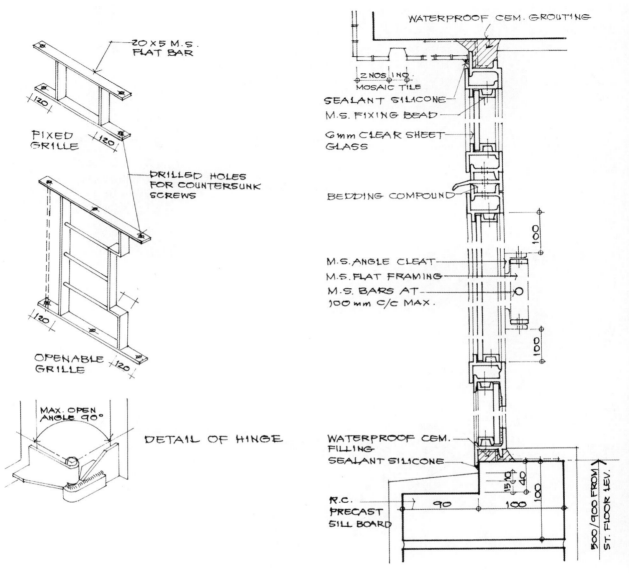

Details for a window frame

Internal modular partition walls

Partition, cooking bench, and sink units

Standard modules form the partitions of the flat. Sizes of 300, 450 or 600 mm by the room height with 75 or 100 mm thickness suit the different locations. These are constructed with blockwork but can also be built by drywall, which has the advantage of easy assembly or disassembly to give a flexible layout to the flat. The material for drywall is relatively light and made of cement and wood fibres, without any concreting process required for erection.

Cooking benches are prefabricated elements. The bench top is cladded with one piece of 0.8 mm satin finish stainless steel sheet. The bench itself is lightweight concrete or similar material to the drywall system and supported by 60 mm concrete support, smooth finish, and epoxy painted.

The stainless steel sink is cast in a lightweight concrete bed, 75 mm thick, to afford strength to stand chopping and cutting. Earthing lugs are designed and built for both sink and bench.

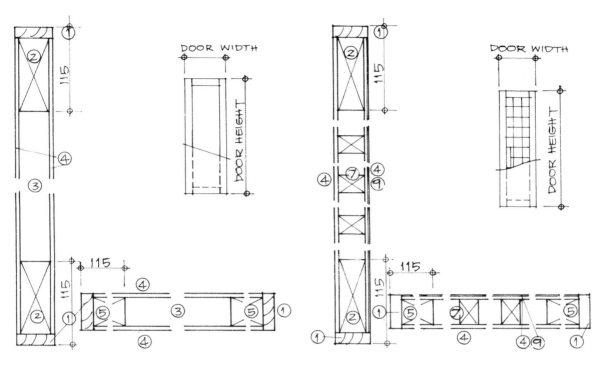
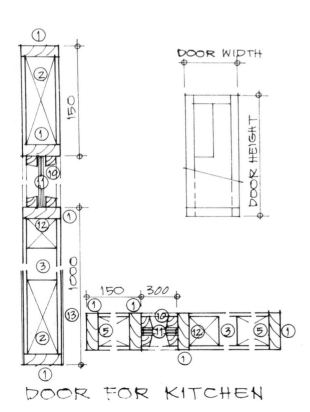

① 15mm TEAK LIPPING
② 40mm H.W. RAIL
③ H.W. CORE TO G.S.10.54
④ 5mm PLYWOOD FACING
⑤ 40mm H.W. STYLE
⑥ 35mm H.W. RAIL
⑦ 35 × 25 SOFTWOOD EGGCRATE RIBS AT 150 c/c B/S
⑧ 35mm H.W. STYLE
⑨ 1.3 PLASTIC LAMINATE AT INNER SIDE
⑩ 15×15 TEAK BEAD
⑪ 6mm GEORGIAN WIRED POLISHED PLATE GLAZING
⑫ 40 × 40 H.W. FRAME
⑬ 1.6mm × 150 H. ALUM./S.S. KICKING PLATE (IF SPECIFY)

Door details for a kitchen

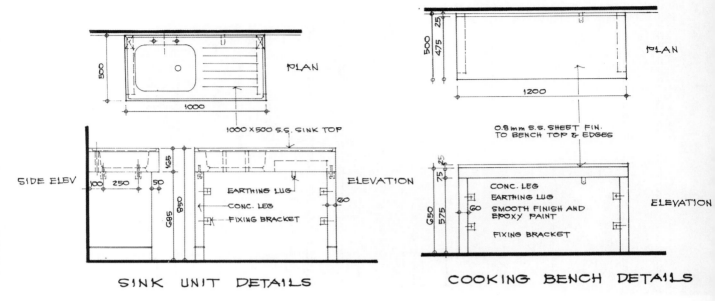

A kitchen unit

Completion of finishing and windows for the prefab façade

砌牆及廚房的設計

如何把模組設計的概念推廣，並得到從外至內的效果，是需要很多因素的配合。細緻如廚房的鋅盆位置和煮食的案頭，都是以劃一的規格為設計標準的。

Quality improvement

The Housing Authority has sought quality assurance from design through manufacture to delivery and storage on site.

The quality-assurance system is sought through a third-party technical check and ISO 9000 standards. PSA International from the Department of the Environment UK was engaged for an independent technical check, including review of production drawings, interviews with the project team, and visits to mock-ups and construction sites. The ISO 9000 series from the International Standards Organization set out how to establish, document, and maintain an effective quality management system.

A Performance Assessment Scoring System (PASS) is an objective performance assessment of individual contractors. Four general aspects are assessed covering the management input assessment, programme and progress assessment, environmental and other obligations assessment, and safety assessment.

Contractors with excellent performance are awarded greater tender opportunities, whereas those failing to comply with specifications may be blacklisted. An approved list of building contractors for the Housing Authority is being established with the launch of the new Harmony Blocks, aiming at assured quality and better living.

提高質素的設計程序

一直以來，香港房屋協會為市民提供高質素的樓宇居所。對於品質的管制尤其嚴格。

質量管制的指標是按照下列各項國際準則訂定的：

1. Third-Party Technical Check: 第三者的技術驗查
2. ISO 9000 Standards: 提供有關品質管理的指引
3. PSA International: 它的指標是來自英國的環境保護局。除了審查施工圖則外，還協調工程部以了解工程的實際進度。
4. PASS (Performance Assessment Scoring System): 這是一個監察承建商工程表現的制度，主要分為四方面：管理能力評核、工程計劃和進度評核、環境和其他責任的評核和安全評核。

總之，從紙上的設計草圖開始，到施工期間的地盤視察、抽樣檢驗、入伙驗樓等一切事項，都得到充份的監管。

Architect and Engineer:	The Hong Kong Housing Authority
Contractor:	Shui On Construction Company Limited
Year of Completion:	1992

3.5
Residences at Nos. 96, 98, and 100, Ma Ling Path, Sha Tin: Single-Storey House Construction
沙田馬鈴徑洋房：小型獨立式住宅建築

Concrete shell

The three houses are single-storey residences, planned for a family, with a large garden and lawn, a yard and carpark. Architecturally, the three houses are designed with the idea of playfulness through the use of geometrical forms and the cheerful primary colours. The prismatic triangular roofs cap each house, symbolizing the hilltop on which these houses are sited.

坐落於沙田馬鈴徑上的三間小平房，有著趣味性的組合設計。純白的主體，穿插著色彩鮮豔的立體幾何拼合。驟眼看來，頗有點像小時候愛玩的積木遊戲，三角形的屋頂覆蓋，再配合周圍的山嶺，這一切刻意的組合，充份反映著設計者的一番心思。

Structure and ground floor

Completed in 1986, this project is an example of a single-storey house construction. The basic structure is 150 mm thick reinforced load-bearing walls on strip footings. The ground slab rests on grade with PVC sheet, 75 mm binding layer, and 100 mm hard core. Additional waterproof cement sand screed on bituminous paint was added on external wall with adjacent planters.

結構及首層地台

這三間小平房，在 1986 年落成。因為建築物坐落在平地之上，所以，它的建築結構也比較簡單。主要的外牆為 150 毫米厚的混凝土，配合條形基腳，首層的混凝土底板以地面承托，100 毫米的硬核底之上是 75 毫米的黏合層，最後再加上防水的膠布。至於外牆混凝土，除了用防水英泥外，還多加一層瀝青油作保護。

Skyline shaped by the form of the building

Elevation facing a garden

Elevation facing an entrance

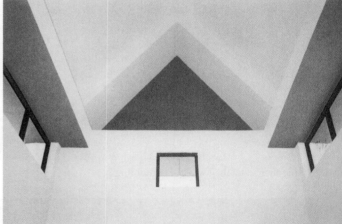

The triangular roof: external and internal

The three houses amid the surrounding landscape

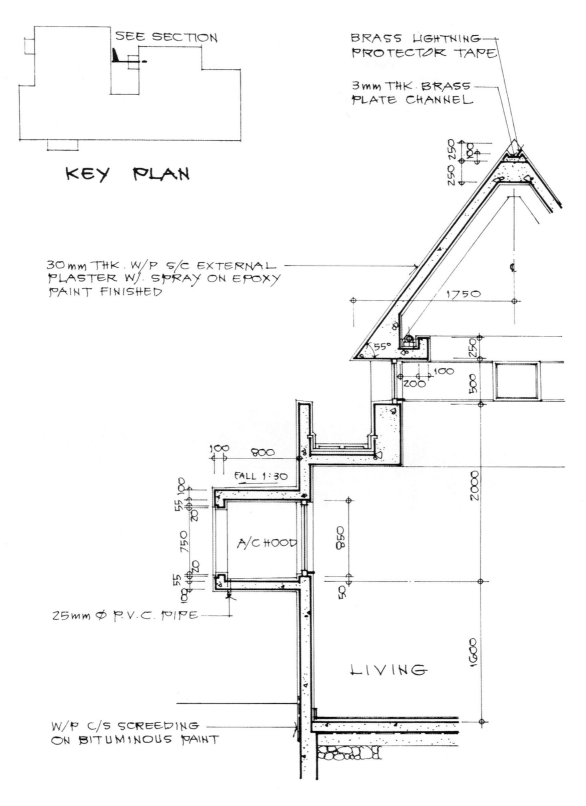

A wall section

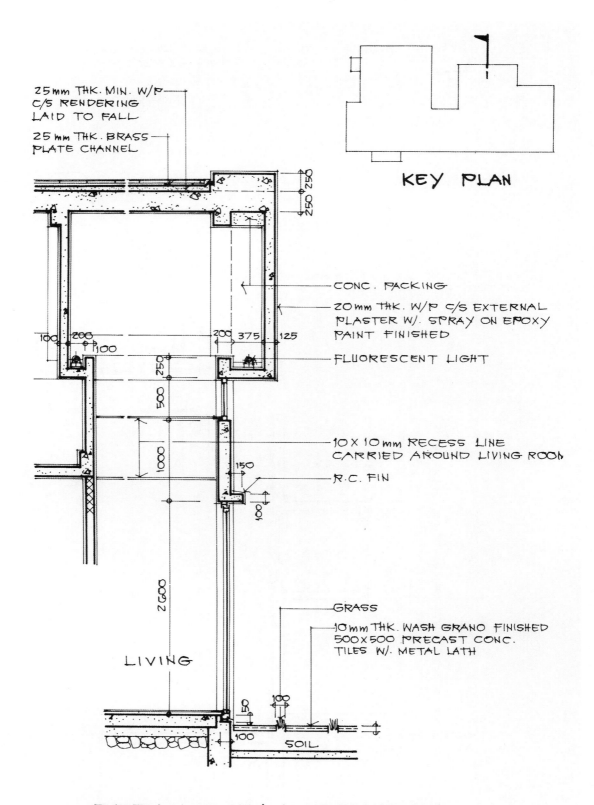

A wall section showing roof

Pitched roof

The pitched roof is inclined at an angle of 55 degrees. Because of the waterproof cement sand external plaster and epoxy paint finish on the 125 mm thick concrete roof, there is no report of water seepage. The roof ridge is fitted with a brass lightning protection rod. A brass plate channel is installed to collect all rust from the brass rod and tape, to avoid stains on the roof. An inverted concrete light trough is formed around the inside of the pitched roof to produce an artificial concealed lighting effect.

三角形的屋頂設計

125 毫米厚的三角形屋頂，與水平成 55 度角。周邊塗上防水的英泥和乳劑油，至今未見有漏水的情況。

　　三角形的尖頂部份藏著 U 型的避雷帶。此外，U 型的管道可以把避雷帶的銹匯集，而避免銹漬染污外牆。內部設計也見心思，三角形屋頂的底部設置向上燈槽，加添一份別緻的氣氛。

Architectural features

Concrete housings have been designed for the air-conditioning hood and the overflow pipe for the flat part of the roof and appear as geometrical forms of architectural features. The main entrance door is designed with a grid pattern of a teakwood frame glass door, which echoes the grid of recess lines on the building façade. Access for the drainpipes is by means of concealed panels on the external wall surface.

A sample of a wall finish

Setting out of paving

Chequerboard bays of concreting of paving

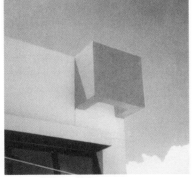

Rainwater overflow and entrance door

Blockwork fence wall

Case Studies

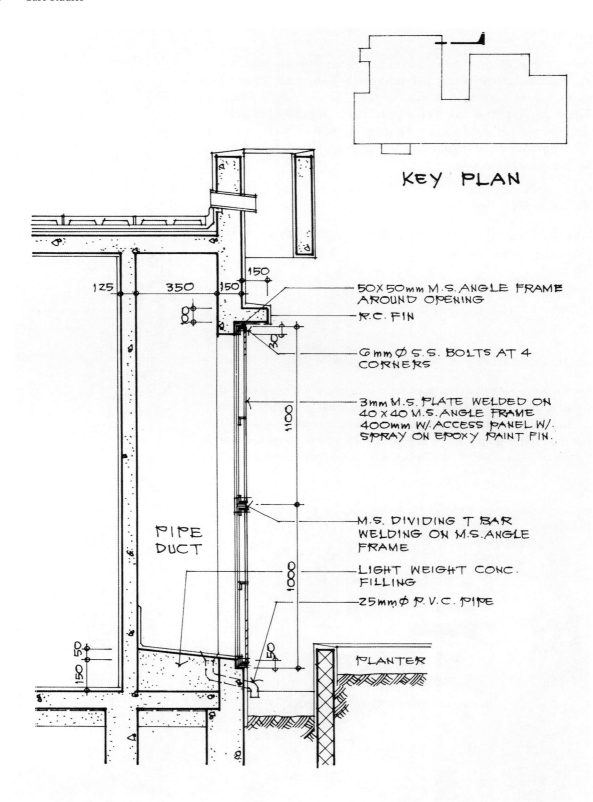

A typical wall section showing the rainwater overflow from the roof

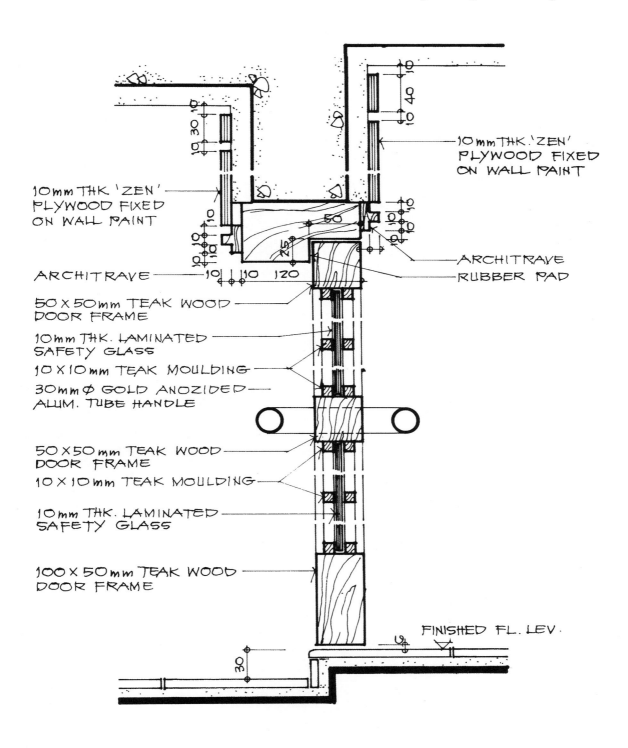

Entrance door details

建築的裝飾

成功的設計，是運用建築的裝飾來遮蔽該要暴露的外牆設備喉管和雨水管道。那些似隨意附加的立體幾何裝飾，原來是作掩飾喉管的用途。雨水疏喉收藏在外部的暗牆處，只要開啟一扇活門，便可以清理及維修。

　　正門的設計以方格組合為主體，基本的圖案和設計的大小，亦與樓身的模數配合，使之為劃一的設計概念。

Client:	Central Management Limited
Architect:	Ng Chun Man & Associates
Structural Engineer:	Architects & Engineers (HK) Ltd.
Contractor:	Yu Hsin Construction Limited
Year of Completion:	1986

3.6
Mong Tung Wan Youth Hostel: Small Building Construction
望東灣青年旅舍：小型建築建設技術

Though this is a small project with a modest budget, the architect put much effort into investigating different aspects, from conception and manipulation of spaces to detail, texture, and proportion. To start with, the architect had to understand the Youth Hostel requirement and the site context. The Hong Kong Youth Hostel as a charitable organization provides accommodation and opportunity for youth to explore the surrounding countryside. The site, Mong Tung Wan, is a small sheltered bay on the south coast of Lantau. It is against this picturesque setting of gently sloping terrain that the Youth Hostel as a sensitively designed environment was designed and built.

香港青年旅舍是一個非牟利的機構，專為青年人提供一人郊外住宿的地方，讓他們有更多機會接觸大自然。

而位於大嶼山南岸的望東灣青年旅舍，也許就是眾多青年旅舍設計的表表者。建築設計師需要了解旅舍的運作、望東灣的地理環境等，把豐富的建築意念，完全呈現於空間的結構組織、建築的細部、比例與質感之中。建築物的編排與佈置，在望東灣的山嶺之下，構成了一幅絕世的圖畫。

Basic design concept

The solution to this particular setting is a cluster of interlinking and distinct blocks containing different functions corresponding to the scale of country village houses. Steep pitched roofs make reference to the sloping terrain. A stepped terrace at the central court of the cluster is conceived as an external 'room' embraced by the building façades.

基本的設計概念

建築物由幾個不同功用的小屋組成，形態和意念來自鄉間小屋。斜斜的屋頂和背後的山嶺，形成了一個有趣味的呼應效果。小屋與小屋之間組成的中庭，仿似承接著從上而來的勢力，如此的庭子也構成一個很重要的空間象徵概念，庭為「空」，但被「實」的建築群主體環抱著。

Mong Tung Wan Youth Hostel

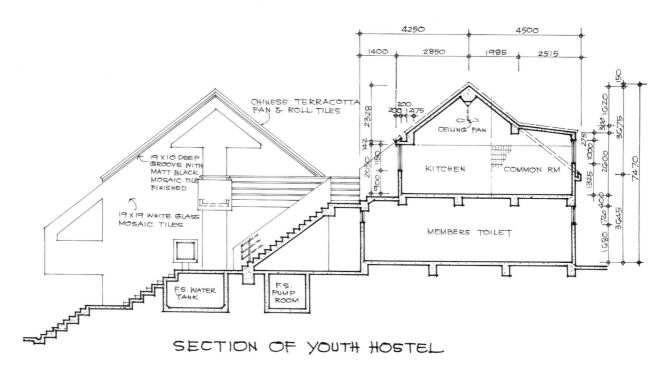

Section of the Mong Tung Wan Youth Hostel

Architectural details

The buildings are built on foundations of strip footings founded on the bouldery colluviums. The 200 mm thick load-bearing walls support the main roof slabs constructed at a pitch of 40 degrees. Red-brown clay tiles are laid on overlapping joints covering the roof. The rainwater gutter is concealed within the pitched roof profile.

Due to intense usage of the youth hostel, windows are made of heavy-duty, hot-dip galvanized mild steel sections with Georgian-wired glass. All finishing materials and fittings were selected for their durability and ability to stand hard wear. The internal floors and balconies are paved with heavy-duty quarry tiles. Galvanized iron pipes are used as railings. The spiral staircase consists of prefabricated step units of cast iron.

External landscape details are treated in a poetic way, such as combining natural boulders with the geometrical planters. The triangle as the logo of YHA is echoed in many forms of details—shape of window openings, roof forms, symbolic logo in galvanized iron pipes—all responding to the beauty of the site and the freedom of hostel life.

建築細部

建築地基採用條形基礎，40 度斜角的三角形屋頂則由兩幅 200 毫米厚的承重牆支撐著。三角形屋頂之上，由泥紅色磚瓦相疊而成。疏水管則收藏在三角形的屋頂裏，同時達到實用和美觀的效果。

青年旅舍為一公眾設施的建築，所以在物料的選擇方面也較為小心謹慎。除了考慮到建築成本外，物料的耐用功能也自然為主要考慮因素。比如，所有的窗門都採用負重的鉛水片鋼。室內和露天的地台階，

都是選擇那些適合經常使用的地磚；而特別設計的螺旋形樓梯，也是選用鉛水鋼鐵的預製梯形模件的。

　　至於建築物坐落的周圍環境，更是經過建築師的經營和佈置。除了要配合建築物本身的設計外，更須配合現場的環境，達到與大自然融為一體的意義。在戶外園林設計方面，保留了原有的石群，與刻意設計的三角幾何花園作對比。在門樓之前更有一塊三角形的碑牌，正好作為建築物的指引。建築與自然融合，為人們提供一個簡單、詩意的休憩地方。

Client:	Hong Kong Youth Hostels Association
Architect:	Rocco Design Partners
Structural Engineer:	Peter Pun & Associates
Contractor:	Condor Construction Co. Ltd.
Year of Completion:	1983

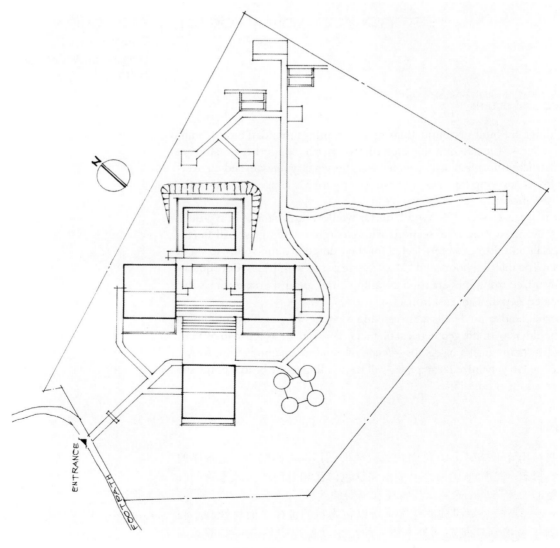

Master layout plan

Tiled roof

Concrete portal

Balcony with BMS railing

Steel windows

Barbecue pit and bench

Concrete seats and planter

Central court

Steel spiral staircase

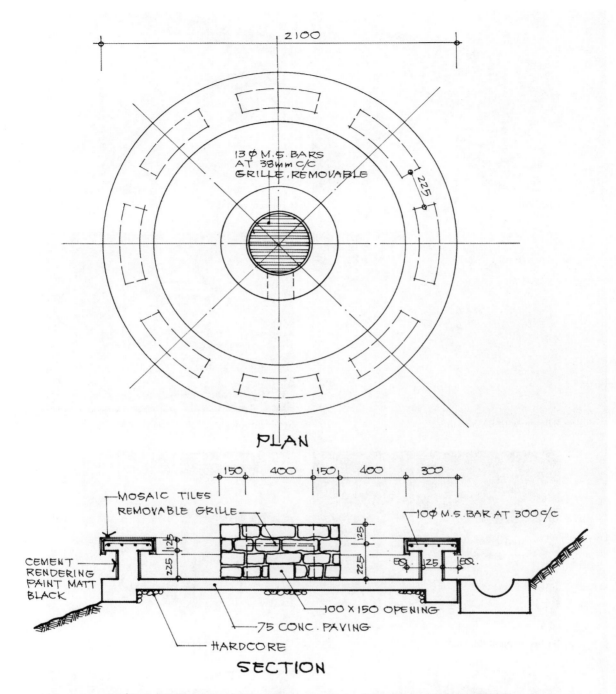

Barbecue pit and bench details

Articulation of small buildings

3.7
Birchwood Place, 96 MacDonnell Road: High-Rise Residential Building on Slope
麥當奴道96號寶樺臺：斜坡上的高廈建設

The completed Birchwood Place

Birchwood Place is one of the 'high-class' residential apartment buildings in Hong Kong. There are some unconventional techniques involved during construction that are listed for discussion. The difficult site conditions, special structural requirements, as well as architectural design for function and aesthetics contribute to make this case study unique.

Located between Bowen Road and MacDonnell Road, the site is in fact a steep slope with a 20 m level difference. This posed difficulty in foundation and site formation design and construction work.

Thirty-eight residential floors stand on a transfer plate with ground floor and five levels of a semi-basement carpark below. The remarkable point is the 'curve-cornered' triangular plan forms housing three domestic flats to command views of the harbour. The result also provides a lack of a 'back elevation', an unsightly view so common with the 'aeroplane-shaped' or 'T-shaped' residential plan of conventional Hong Kong residential blocks.

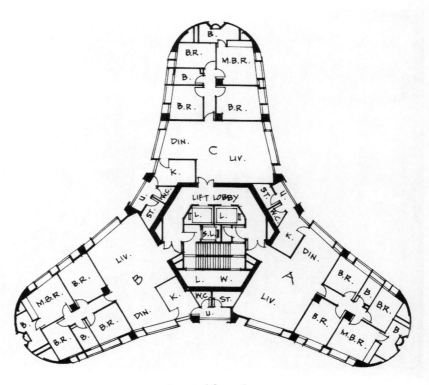

A typical floor plan

一般高層樓宇建築比較困難的地方，是要解決結構設計的同時，還要評估它對於建築外觀造成的影響。

但對於這幢樓高 38 層，位於麥當奴道與寶雲道之間的高尚住宅大廈而言，除了本身的結構設計之外，更須考慮到這段半山上的斜坡為工程施工所帶來的困難。

地盤本身是相差 20 米高度的山坡，在地基設計和建築施工上，會有一定程度的困難和問題。此外，此幢大廈的設計，打破以往住宅樓宇中經常出現的（飛機式）或（T-式）的編排，大廈的外型成自然弧形的三角組合，每層的三戶住宅單位都可以避免有互相對望的情況，而且每個單位都擁有望海的景觀。獨特的外型之餘，更有實際的效用，實為建築設計中難得的例子。

Caisson foundation

After demolition, construction started with the digging of a caisson foundation and caisson walls. The tower is resting on a caisson foundation with the central core, also constructed by the interlocking caisson method below ground level. These foundation caissons are about 30 m deep, the last 10 m penetrating into solid rock, and are 1.5 m to 2 m in diameter. In the actual construction, work progress for caissons through soil was about 1 m depth per day, whereas that through rock was only 0.3 m depth per day. Work was eventually accelerated by blasting, with approval sought from the Mines Department.

沉箱的地基

大廈的沉箱地基直徑長 1.5–2 米，深 30 米。中間的主要核心亦以沉箱作為結構牆。

據高廈的地基結構計算需要，30 米深的地基，當中需要有 10 米深入地下的石之中。這樣艱巨的地盤工程，工作進度仍要維持在每天挖泥 1 米深和陷入 0.3 米深的石。經過礦務部的特別批准，承擔工程的承建商才可以採用爆石的方法，使工程加速完成。

Excavation of caissons by blasting

204 Case Studies

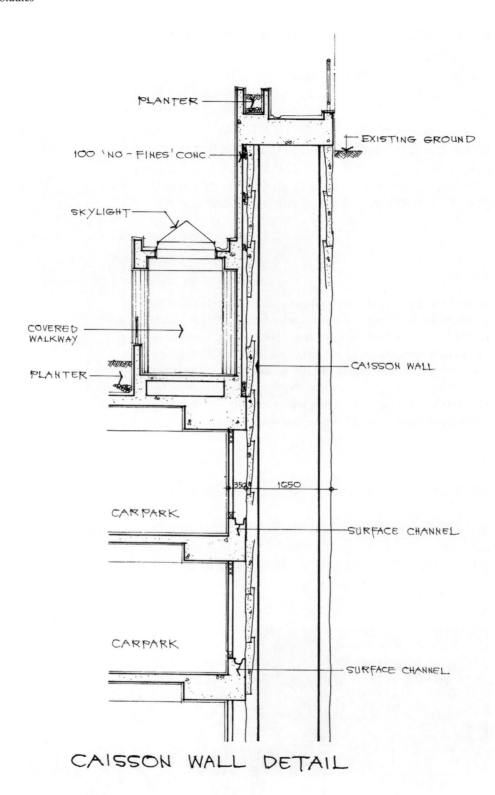

Caisson wall details

Site formation

The caisson wall with some pipe pile walls acted as retaining structures 20 m high for construction of the five levels of carpark in the semi-basement. These caisson walls and pipe pile walls were temporarily shored against the central interlocking caisson core wall. These temporary steel shorings were removed layer by layer as the podium carpark levels were constructed; these act as permanent shoring or strutting after construction.

地盤的準備

以沉箱及管樁為護土牆，承擔了原來地後的 20 米之間距離的泥土壓力，使之作為五層半露天停車場之用。興建時，旁邊用鋼支架作臨時護土支撐。隨著每層停車場的建成，臨時的鋼板支撐逐一拆去，直至沉箱牆完成之後，成為永久的護土牆。

Transfer plate

The typical floor columns were changed into fewer but larger columns at the ground level through a 3 m transfer plate. As the ground level is not on grade, temporary propping for the heavy transfer plate had to be carried through several floors. But the contractor devised a way to construct a strong slab for the transfer plate and used that as part of the strutting for the transfer plate.

承重層的結構

地面上一層的 3 米厚混凝土層，把以上 38 層的承重結構柱化繁為整，是一種高廈結構的設計模樣。

　　要建築一層 3 米厚混凝土承重結構，是需要堅固的臨時支撐架構，承建商在開始這層結構層之前，預先完成堅固的混凝土柱和牆，作為臨時支撐架構的承重點。

Typical floor

To gain a speed of five days per floor for the concreting cycle, a typical floor was divided into three wings for concreting. The bay windows and other architectural features were not constructed during the regular concreting cycle but later. This was scheduled so that commencement of work for lift shafts and other interior work would not be delayed.

標準層的工程

對於承建商和發展商來說，時間就如金錢般重要。經過計算和考慮，承建商需要維持平均五天完成一層的工程進度。於是，順應著設計的本來特點，把每一層的工程分為三個相同的部份；輪流進行釘板、拆板等建築程序，相同的工序根據編定的時間表進行。

　　至於非結構的建築細部，如窗台、外牆的裝飾架等，為避免影響主體的進度，則安排在稍後的時間裝嵌。

Transfer plate at the podium

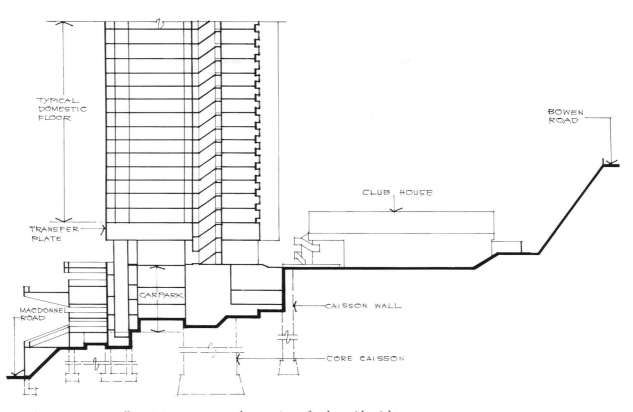

Section showing caisson wall retaining structure and core caisson for the residential tower

Completion of the transfer plate and the commencement of a typical floor

Shoring against the core caissons and the caisson wall at the back

208 Case Studies

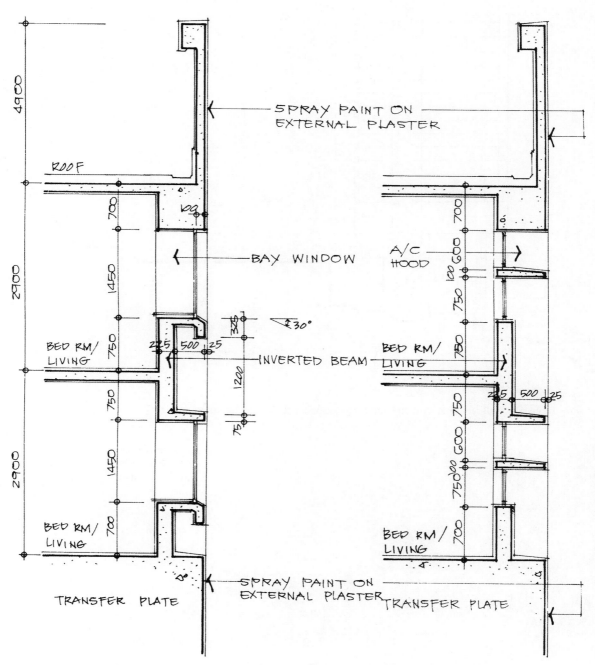

Typical wall section

Wall tiles are carried down to the ground ceiling

Elevation with bay windows and air-conditioning hood

External tile

To overcome the movement problem for this tall building, the external tiles were fixed with special adhesive, allowing flexibility in movement. Movement joints were also provided at a 3 × 3 m grid to cater for expansion. Black and white were chosen for the external glazed tiles. Pointing was used for the final finish at the tile joints.

外牆裝飾

考慮外牆裝飾時，須顧及不同室溫對外露物料所造成的膨脹與收縮效果。故此，外牆3×3米的方格線，除了有美觀的效果外，還可預留作外牆紙皮石裝飾的「活動」之用。

Elevation with bay windows and air-conditioning hood

Swimming pool

The swimming pool is constructed on an elevated structure. A basic reinforced concrete structure provides the external shell for the pool. Asphalt tanking was then applied for waterproofing, after which another reinforced concrete shell forms the internal layer where the pool tiles were laid. A screen channel system is used for filtration.

大廈設施：泳池的設計

為符合高尚住宅的要求，私人的康樂設備自然不可忽略。寶樺臺的設計中也附設了私人會所、泳池等一般的設施。

露天泳池設置在會所的頂部。基本的混凝土結構成為泳池的牆，再塗上 20 毫米石瀝青在主要的牆內，最後加建 100 毫米厚的內牆，加強泳池的防水保護性能後，才鋪上游泳池的紙皮石。

Access road

As access to the site is on MacDonnell Road, an elevated access road was built to go up through the five levels of the carpark to the ground floor. The road surface is finished with 75 mm thick wearing slabs with reinforcement mesh and waterproofing admixture placed on the structural slab. The entrance has to satisfy the sightline for the safety of vehicular traffic.

車道入口的編排設計

從麥當奴道到達大廈的主要入口，要通過一條新建的車路，再途經五層的停車場，才來到真正的大門口。這條新建車路的標準，必須符合政府路政署的最低要求：除了基本的結構外，還要達到防水及耐用的效能。汽車入口亦要適合架車人士的視線，以策安全。

Developer:	Maranta Estates Limited
Owner:	Great Eagle Estate Agents Ltd.
Architect and Structural Engineer:	Ng Chun Man & Associates
Mechanical and Electrical Consultant:	Associated Consulting Engineers
Geotechnical Consultant:	Maunsell Geotechnical Service Ltd.
Main contractor for site formation, caisson foundation, and superstructure:	Tak Son Engineering Co. Ltd.
Year of Completion:	1989

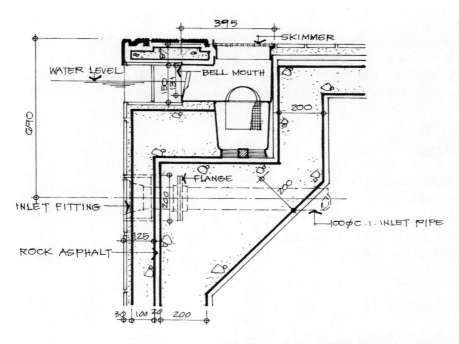

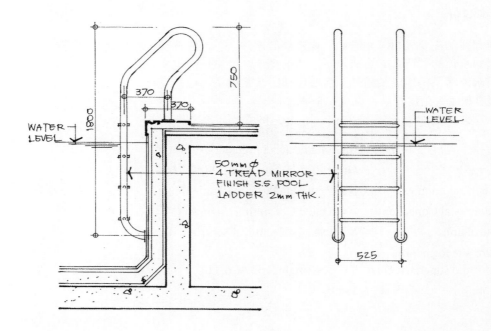

Detailed section of the swimming pool

Details of the swimming pool

Car park entrance

3.8
The French International School: External Wall and Auditorium
法國國際學校：外牆與禮堂建設

Façade of classrooms

To find a design solution that could satisfy the identity of an education system for the French community, to make use of the sloping site, and to conform to the four-storey height restriction, the architect put all the building blocks together in the form of a pinwheel generated from a sky lit central court designed as the focal point and as the place for orientation of direction to the four wings of the primary school, secondary school, administrative block, and recreational block, all interconnected by split-level open corridors.

By placing the blocks at different levels, the sloping site is fully utilized. Viewers can enjoy good scenery from almost every corner of the blocks. Even the auditorium is located in the basement to make full use of the sloping site.

The project was completed in 1983. Later, in 1991, a four-storey block of 16 classrooms was constructed as an addition to the school, with similar architectural details.

在山坡上的一個小地段，依循政府條例對學校設計的高度限制，同時要從設計中反映法國的教育特徵，確實是項不容易的建築工程。

建築師利用中庭作為建築物的重心。繞著中庭為幾座不同功用的建築物，如小學部份、中學部份、行政大樓和康樂大樓等。平面的圖形看來，有點像風車的模樣，是一個充滿動感的構思。

因為地段的斜坡問題，所以在設計上，建築師巧妙地把不同用處的建築物，配合不同的設計水平面高度，以露天的走廊把它們互相連接起來。如此的設計安排，消除了斜坡對於建築的限制。而且在建築物的每一個角落，都可享受到周圍的景緻。

Entrance of the school

Building set on a sloped site

216 Case Studies

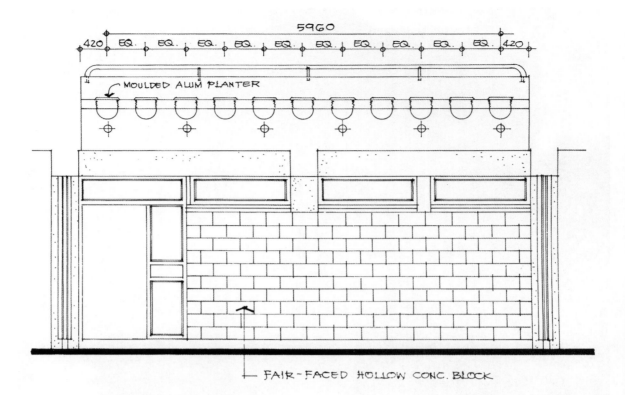

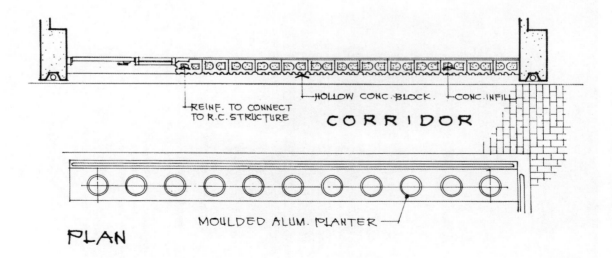

Typical details along the corridor

Central circulation courtyard

Façade of classrooms

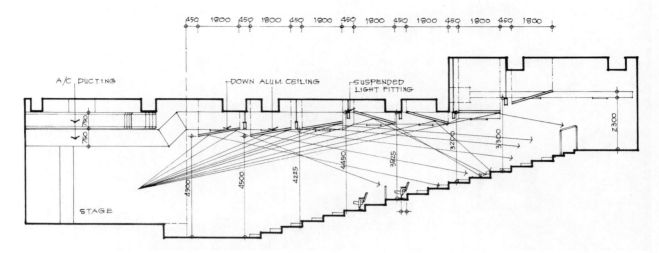

Visual angle is considered at the auditorium

Architectural details

The rugged texture of conventional fair-faced hollow concrete blocks forming partition walls for the classrooms are expressed as a new life on the corridor. However, the concrete blocks inside the classroom are plastered and emulsion painted to give a more neutral texture for the classroom.

Drainpipes were designed with the columns to form an architectural element that expresses truth of construction and services. Inverted dome-shaped MacLeod aluminium planters were added along the corridor to add colour and life to the school environment.

Facing the outside wall of the classroom, fan coil units were installed; 200 mm diameter air louvres were installed and expressed on the external wall as a uniform pattern. The auditorium is designed with an acoustical reflective ceiling together with acoustic absorbing panels in the end wall. The latter is treated with fire-resistant paint every year, because the auditorium has to satisfy the local fire code as a place of public entertainment.

Skylights were built on a 1,300 mm grid of reinforced concrete frame for walkways and pergolas. Acrylic domes incorporating surface drains were installed.

Graphically, reds, blues, and whites are used to symbolize French dominance.

建築的設計特點

以未經打磨的傳統空心磚作為課室的間牆，使外面的走廊通道有著一種難以形容的生氣。但課室內，牆身都是經過粉飾後再加灰水英泥，整齊而雅觀，配合課室內的學習氣氛。

此外，倒置的鋁質模製圓拱型花盆圍繞中庭的走廊，亦扮演了很重要的角色，為校園的學習環境添上一份色彩。

建築設計時所忽略的一般建築設備，如雨水喉、空調安置等，在這裏都經過細心研究，成為建築的細部，同時達到外觀與實際的功用。譬如垂直的雨水管收藏於柱的暗角內；而空調用的通風百葉，在外牆上排列出一個統一的形態，使單調的外牆變得有韻律。

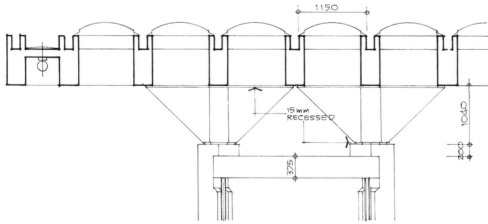

SECTION

Roof skylights with acrylic domes

禮堂是一般學校必需的空間之一。這裏的禮堂設計更是特別，除了安裝有音響用的天花外，還在周圍加上吸音板來控制音響的強弱回聲。根據香港的防火條例，這些吸音板是需要每年都塗上一層防火油的。

天台休憩的地方，興建了一條有蓋的行人廊，當中的設計還得考慮如何擺放雨水喉管和透明的半圓天花。

在平面設計上，建築師還利用紅、白、藍三色去表現法國色彩。

Client:	French International School Association, French Consulate
Architect:	Patrick Lau of Design Consultants
Structural and Geotechnical Engineers:	P. S. Chung & Associates
Building Services:	Mao & Partners
Interior Design:	Ruby Lau
Landscaping:	C. K. Wong
Quantity Surveyor:	Eric Cheng
Acoustics:	Stephen Lau
Theatre Lighting:	Ian Campbell
Theatre Sound:	Jacek Figwer
Completion:	1983–91 (additional Phase)

External and internal walls of classroom and concrete block details

3.9
The Heungs' Residence at the Peak: Granite and Glass Technology
山頂香氏大宅：麻石與玻璃裝置技術

The Heungs' residence

This residence on the Peak was tailor-made to meet the client's specific needs. However, there was still much room left for the imagination of the architect. To address the airiness and open nature of the site context, an interlocking double volume of space was created inside the three levels of the residence. A constantly changing spatial quality is felt for the actual experience through the residence. The Italian interior designer, IMA, enhanced the flowing space quality by curvilinear inlaid marble on the floor and lacquered wood panels on the ceiling. Glass and granite give a textural contrast on the external surface.

坐落在山頂道的這幢豪華私人住宅，帶著一種令人難以想像的雄偉氣派，正好是為它的主人香先生度身訂造的設計。在整個設計和構思的過程中，建築師亦絞盡不少腦汁，把屋主的夢想境界變成真實的樂園。

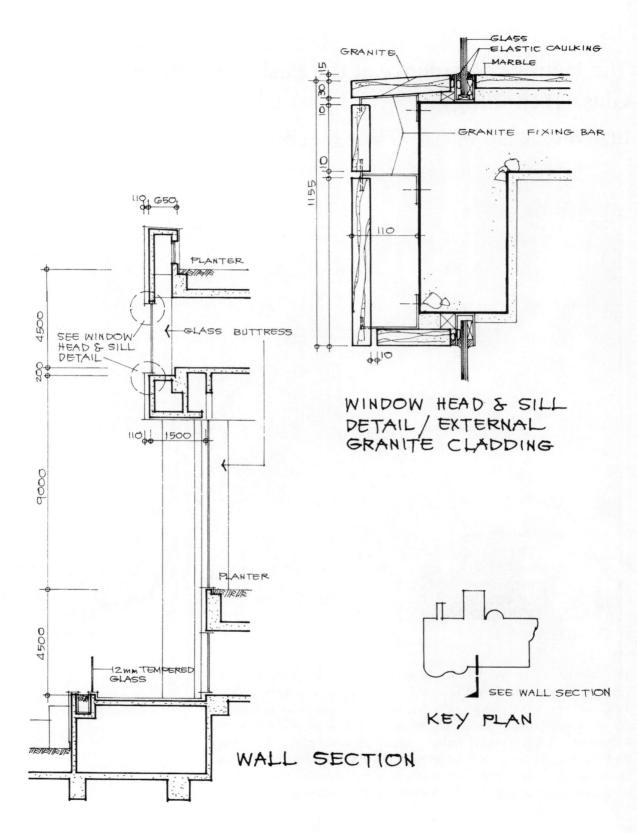

Window details

Construction photo showing the superstructure

　　地段的位置居高臨下，四野曠闊。取其地理環境的優點，設計也盡量反映空間的廣度和深度，使居住者更感與自然的結合和開敞舒適的生活。住宅共分三層，當中以一兩層高和向海的客廳為主體空間，緊緊連接著的還有很多趣味與氣氛不盡相同的大小房間。變換的空間構思，加上由意大利室內設計公司 IMA 的精細粉飾和設計，廳堂的地板配上美輪美奐的意大利雲石、木刻裝飾的吊層天花等，都豐富了居住者的生活情趣和感受。

　　然而，香氏大宅的外圍園林設計和麻石的鑲嵌技巧，也為整幢建築物帶來不少獨特格調的氣派。

General construction

The soil condition is not very good, so caissons up to 30 m deep and with bell-out at the bottom were used as the foundation. Site investigation was done along with the caisson work, to determine the acceptable depth.

When Hien Lee, the main contractor, took over the site, each caisson was connected to individual columns for the superstructure. The first decking was the most difficult, as steel propping had to be used to stand the first concrete slab on the slope. To meet the accuracy of dimensions for the fine interior furnishing and external granite work, special formwork was used for concreting, and careful setting-out work was required. Site coordination with the granite and glass interphases was not easy.

工 程

經探土報告的結果，地基工程只能採用 30 米深的沉箱方式，而且沉箱的底部更要略為加闊，使地基更堅固。

地基工程完成後，上蓋工程的承建商先把每枝的地基沉箱連結成柱身，才繼續以上的結構。首層的混凝土平台最為困難，原因是支撐的頂柱只能負力在斜坡之上。因此，它的建築時間亦比較長。興建時，為著配合精確的戶外麻石及室內雲石安裝，建築師要特別注意地盤的度尺和平水點位置。安置麻石和大型玻璃的工程亦需妥善協調。

Granite technology

Granite quarried in Korea was selected as the basic external finish after an impressive visit to the Sekigahara stone factory in Japan in 1986. Both flamed and polished granite were used to give a subtle contrast to certain surfaces. The architect believed that a polished finish would be more appropriate for the project. However, the client preferred a flamed finish, which gives a lighter colour and needs more frequent maintenance.

The general fixing method is dry-mount and open-joint. The 30 mm thick granite panels are fixed by stainless steel anchor bolts with 80 mm clearance between the concrete surface. The typical panel size is 1,115 × 1,115 mm. A 10 mm tolerance has been allowed for the concrete structure.

To cope with the building regulations, a structural calculation for the granite anchors had to be submitted for approval. The use of the open joint system gives a neat and solid appearance to the granite panels. Joint width is 10 mm, and no sealant was applied except at the roof coping and around the windows. The advantage is that water will drain out easily and will not collect inside the panels. However, the waterproofing barrier was placed on the concrete wall, which is 150 mm thick. Further external waterproofing coating was applied on the finished concrete surface. Also, to minimize the concrete tolerance, special formwork (Visa-form) was used during concreting to give a straight and smooth surface.

麻石的安裝技術

1986 年，業主和建築師應 Sekigahara 石廠的邀請，同赴日本參觀麻石廠。及後，業主決定外牆以由韓國礦洞所開採的天然麻石為主。

若以單一的、純白的麻石為主體設計建築物的外觀，會比較單調。因此，石面運用了火焰燒和拋光製造兩種不同的效果。前者需要經常清潔保養。

這種麻石安裝方法為開口的乾式裝置。把每塊 1,115×1,115 毫米的 30 毫米厚石塊以錨栓安裝在距離 80 毫米混凝土牆之上，結構計算和錨栓負荷程度，都要得到屋宇署的批核和同意才可進行。

至於留空接口式的麻石安裝方法，除了外觀特別之外，所有的雨水或凝結水都在內部的錨栓之間流走。不過，由於混凝土牆較容易接觸水氣，外牆更加厚至 150 毫米及塗上防水油。

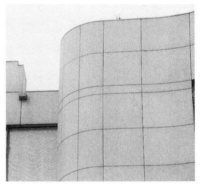

Granite details of various articulations

Granite features

In addition to planar forms, granite can be ordered to be cut as curved forms to match the required geometry of the plan. It is possible to cut these granite columns in a circular shape, which will create an elegant and majestic look but is expensive. Solid granite is suitable for making architectural features such as waterspouts, small posts, and steps, giving a sense of refined artisanry and a feeling of solemnity one can only find in natural stone.

麻石的特徵

細緻的設計融合了麻石的編排與形式。如何保持石塊的特色，充份發揮它原有的浪漫與古典質感，使建築物看起來更為雄偉、典雅，都是一門值得研究的學問。

這樣，建築師在設計時早已考慮到石塊的大小、形狀，於工場加工或切割，然後在實地安裝。特別效果的設計，使門前圓形的柱子、雨水管道、水池等地方成為建築特色。

226 Case Studies

View of the entrance and granite-covered façade

View of the garden in the foreground with granite-covered façade

Lighting effect shown on the building façade

Lighting design

The interior lighting control system is computerized and master-controlled (produced by Tutron). This is linked with the air-conditioning system and curtain system to allow digital control or timer control for each individual space. Thus this building can be termed an 'intelligent' residence.

照明設計

雖然是私人小屋的設計，室外的照明系統也沒有忽視。這間大屋的所有燈飾是由中央電腦系統控制的，在什麼時候亮燈、調節光暗等，都輕而易舉。這系統亦與空調及簾幕連接，變成「智能大宅」。

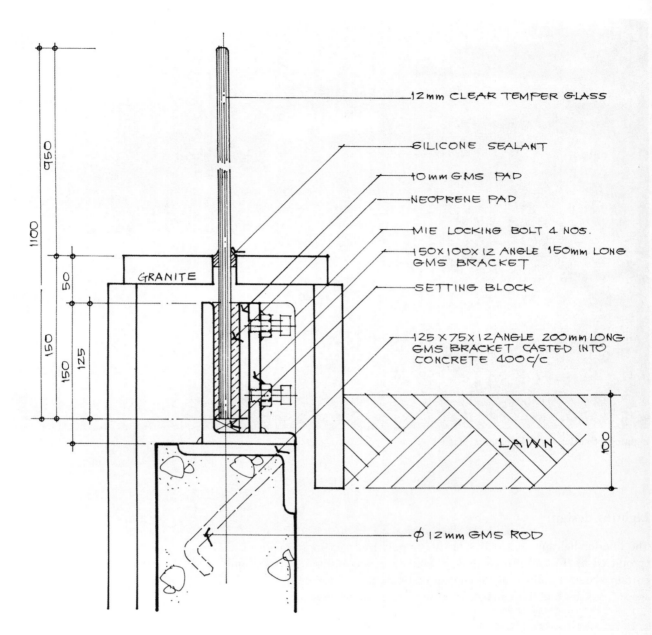

Glass balustrade details

Glass technology

The 4.2 kPa wind load made the design of the 8 m high glass wall in the living room difficult. Asahi glass of a 25 mm thickness was ordered from Japan and erected as a suspension system. Vertical glass fins of 900 mm width are used as stiffeners.

Decks all around the garden are fenced with a 1,100 mm high and 12 mm thick tempered glass parapet. These are designed to stand a horizontal static load of 0.35 KN per m². The porch above the entrance is an interesting structure of stepped, frameless glass parapet roof light formed by supporting horizontal pieces of glass with vertical glass element. Glass blocks are used as a curved enclosure for the internal staircase. Mild steel bar reinforcement is added for structural support.

Design of glass balustrade on the roof

玻璃的運用技巧

綜合而言有四個主要部份：

1. 設計起居室中 8 米高的落地玻璃，先要估算它們所承受的 42 kPa 風力，並不是一件容易的事情。特別設計的懸吊系統，只需要 900 毫米深的玻璃翼、能把每一塊 25 毫米厚由日本訂來的 Asahi 玻璃垂直懸吊。
2. 以玻璃為欄杆，在花園平台上亦成為一個獨特的設計。1,100 毫米高的欄杆，要承受每平方米 0.35 KN 的橫向淨載重，需要 12 毫米厚的強化玻璃。
3. 正門的門廊之上是個有趣的階梯形玻璃天窗。每組橫向和直向的都是 19 毫米的強化玻璃。
4. 玻璃磚用在室內的樓梯平台，半圓形的設計，當中以生鐵作為結構的承重模架。

Living room with 8 m high frameless glass

Case Studies

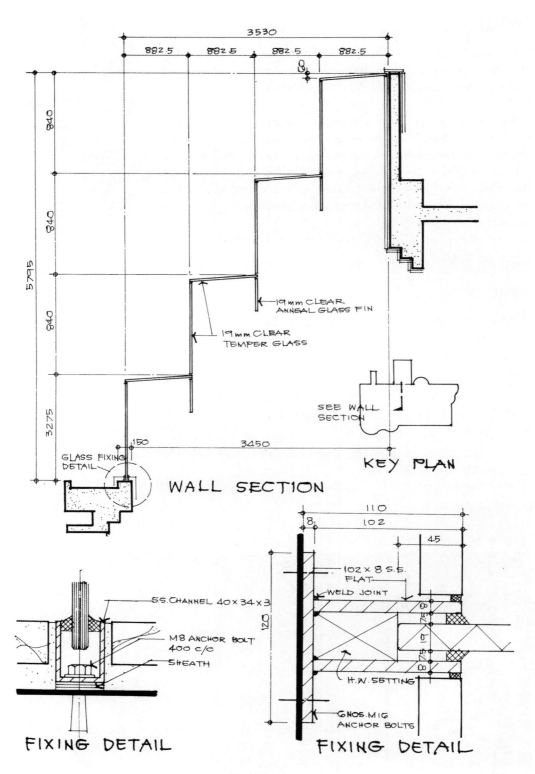

The frameless stepped glazing

Design of the interior

Glass block wall around the staircase

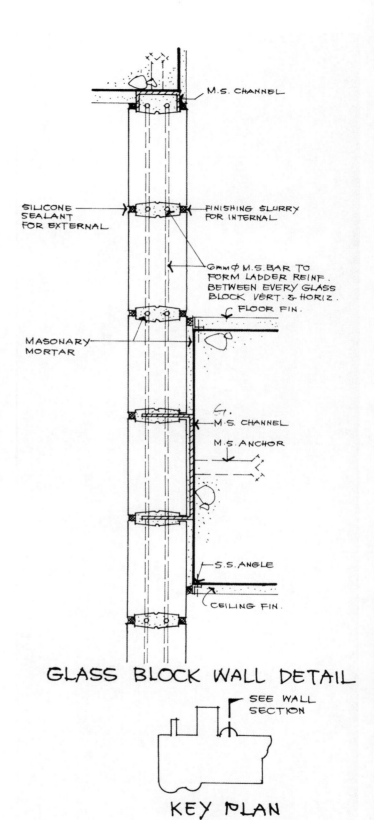

GLASS BLOCK WALL DETAIL

KEY PLAN

Landscape design at the entrance

Client:	Mr and Mrs C. K. Heung
Architect and Structural Engineer:	Ng Chun Man & Associates Architects & Engineers (HK) Ltd.
Geotechnical Consultant:	MVA Engineering Consultants (H.K.) Ltd.
Site Formation Contractor:	Winkey Construction Company
Caisson Contractor:	Yu Hsin Construction Ltd.
Main Contractor:	Hien Lee Engineering Company Ltd.
Granite Contractor:	Sekigahara Stone Co. Ltd.
Interior Designer:	IMA Istituto Mobili Artistici, Italy
Lighting Control System:	Intron
Lighting Consultant:	Stephen Lau of Design Design Architects
Year of Completion:	1990

3.10
The St. John's Building, Garden Road: Curtain Wall and Cladding
花園道聖約翰大廈：玻璃幕牆及覆蓋板

The St. John's Building is designed to relate to the Hong Kong context, blending with the urban setting yet asserting its own identity. The architect, Mr Anthony Ng, believes in the social implications of architecture and created an urban environment around the building, providing both a landmark and a place for the public. The present building plaza was completed in 1983 and has become a focus for people. It thus is a good example of architecture able to stand the test of time.

The stepped platform was conceived as one continuous open space incorporating an exciting water feature as part of the urban landscape. The fountain is designed as a focal point for pedestrian traffic.

At the other end of the site is a freely curved piano-shaped internal courtyard providing a relaxing setting.

每一幢處於煩囂都市的建築物，都扮演著一個重要的角色。摩登的設計，不單要具備耐用功能，還要反映社會不斷的變遷、環境的需要等。

這樣的設計觀念，正好反映在這幢位於中區花園道的聖約翰大廈身上。

由 1983 年建成至今已有頗長的日子，建築物廣場早已成為那裏的一個焦點。門前的台階，接連著繁忙的馬路，成為大廈的正門口；旁邊的小型噴水池，簡單中見特色，且充滿都市的味兒，是討人欣賞的趣味所在。但在大門的這一邊，不規則的流線型中庭設計，也為人提供一處舒適的休憩地方。

Curtain wall and cladding

Rather than an all-glass curtain wall system, a modular cast aluminium cladding system was specified. As the site is a narrow island, exposed on all sides, the cladding system with about 30% glass area will cut down heat absorption tremendously, making the building more energy efficient.

The cast aluminium system was selected after a thorough inspection in the Japanese factory, representing the best product from state-of-the-art technology. The cladding panels have a deep oval texture with a heavy and solid appearance. The panels are finished in grey-coloured polyurethane coating to match the glazing colour.

Each of the standard modules consists of a glazed panel, a spandrel panel, and an internal window trim. The edge of the glazing is finished in a dry-and-wet process with both gasket and sealant. The textured panel is made of 6 mm thick cast aluminium reinforced with cast aluminium ribs around the window opening. The reflective glass is float glass rather than tempered glass, to achieve

Exterior of the St. John's Building

a distortion-free reflection. The glass is 12 mm thick, determined by noise reduction criteria, due to the surrounding traffic noise.

玻璃幕牆與外置牆身

大廈與毗鄰的建築物比較疏離,然而在這個狹窄的地段之上,又可能因為其開敞的客觀因素,而增加建築物本身的吸熱效應。為此,玻璃與鋁質間格牆的比例是整個設計方案的一個基本主題。

要減低大廈的吸熱程度,玻璃的總面積只能佔百分之三十左右;而模數的灰色鋁格間板亦專程在日本訂造。上等的物料為先進科技的產品,配合基本的計算程式,加上橢圓形設計的模數格間板,使大廈的外型別樹一幟。

每模組的外牆板,由一塊玻璃和拱肩鑲板拼合而成。玻璃窗的邊圍用墊料和封料,再經過特別程度密封介面,6毫米厚的鋁板刻有細緻的圖紋,嵌在鋁質的架構中。

慣常採用的強化玻璃,這裏亦派不上用場。取而代之的是12毫米厚的玻璃,除了減低一般出現在強化玻璃的折射現象外,「12毫米」是經過計算之後,還可以降低馬路上嚴重嘈音的影響,使室內的工作環境更舒適。

Client:	Peak Tramways Company Limited
Project Management:	Hong Kong & Shanghai Hotels Limited
Architects and Engineers:	Kwan, Ng, Wong & Associates
Curtain Wall:	Dax Consultants Limited
Landscape Consultant:	Belt Collins & Associates
Main Contractor:	Hip Hing Construction Company Limited
Curtain Wall Contractor:	Tajima
Year of Completion:	1983

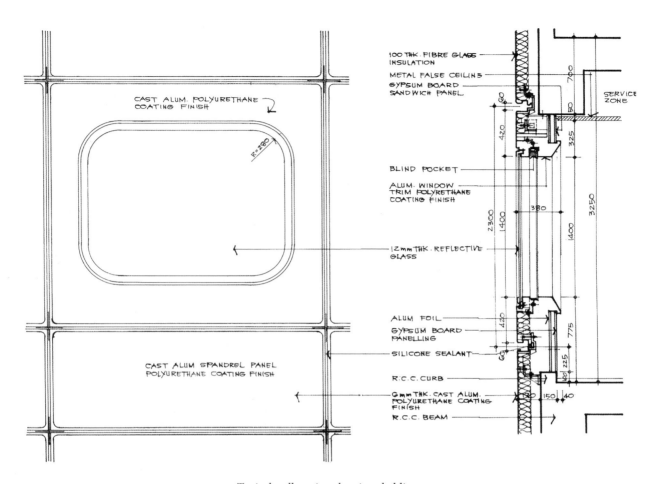

Typical wall section showing cladding

Water feature as part of the urban landscape

The precast aluminium cladding

The internal courtyard

Skylight and the internal courtyard

3.11
Wanchai Indoor Games Hall: Roof Truss Construction
灣仔室內運動場：樑架建築樣式

The Wanchai Indoor Games Hall (now renamed Harbour Road Sports Centre) was designed by the Architectural Services Department (ArchSD), the major role of which is to design and construct government buildings. This games hall commenced tender in May 1985 and was completed in October 1986.

The ArchSD architect works in a multidisciplinary team with the quantity surveyor, the structural engineer, and the building services engineer. Project management duties are carried out by the project team leader, who undertakes the coordination role. This leader need not be a registered architect. The architect controls and coordinates the design and construction aspects of the project. The structural engineer determines the stability of a building, aiming at low budget, and the quantity surveyor gives cost advice and makes contractual arrangements. Lastly, the building services engineer deals with the internal environmental control and utilities provision, such as air-conditioning, light, heating, lifts, and fire-fighting services.

灣仔室內運動場為政府轄下的建築署設計及興建的公眾康樂設施之一。計劃早在 1985 年 5 月構思，直到興建完成並在 1986 年 10 月正式啟用，較一般同類形建築物的建築期為快。

隸屬政府部門的建築署，它的工作範圍相當全面。除了有負責指導及設計工程的建築師外，其他的輔助部門也不少，包括測量、結構工程、機電工程等，使工程能夠順利進行。同時，在工程的效率、速度和品質三方面，亦達到嚴格的標準。

Roof truss

The roof truss is a long span steel structure spanning over a 38.5 m column-free hall for various functions such as basketball, volleyball, or badminton. The galvanized mild steel structure was prefabricated in Japan and assembled by welding and screws on the Hong Kong site. Steel corrugated sheets are incorporated in the double-pitched roof of a 1:10 fall formed by the truss. The contract sum for the truss and roof was HK$3.5 million (1985 level).

Structurally, the truss is based on a 5.5 m ×5.5 m grid, i.e., the main trusses with a depth of 3 m are spaced 5.5 m centre to centre; 1.5 m deep ties cross the main trusses at right angles and at 5.5 m spacing. The 5.5 m grids are further reinforced by diagonal ties, which are also intercepted by 500 mm deep secondary trusses. The main truss is built of 267 mm diameter steel pipes.

Lighting provisions also correspond to the grid with a 600 mm wide catwalk system constructed alongside for maintenance purposes. Basketball goals and dividing nets were also installed and hung from the truss.

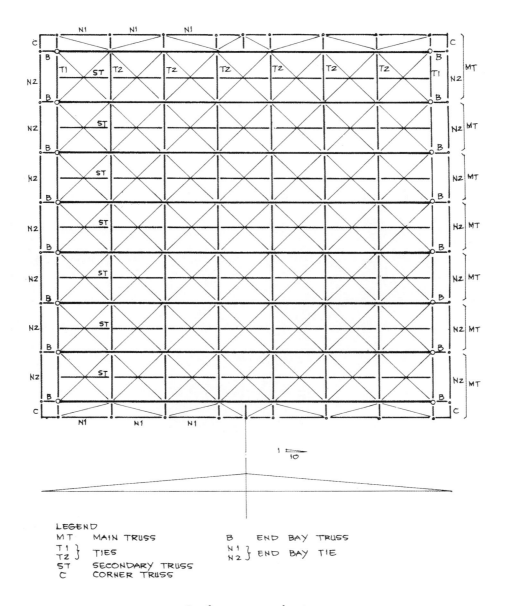

LEGEND
MT MAIN TRUSS
T1 } TIES
T2 }
ST SECONDARY TRUSS
C   CORNER TRUSS
B   END BAY TRUSS
N1 } END BAY TIE
N2 }

Roof truss structural system

The roof

The Wanchai Indoor Games Hall

The roof is covered by corrugated steel sheets with lap joints fixed by self-tapping screws and sealing tape. The end of the truss forming the eaves is covered by metal louvres for ventilation.

天花的樑架

沒有柱的巨型運動禮堂，天花的樑架為一個 38.5×38.5 米、以鋼鐵為主的長跨度結構。能提供寬敞的空間，讓多項球類活動進行。鉛水生鐵的結構板架由日本製造，運抵香港後，才安裝為天花的樑架。屋頂的斜度為 1:10，由兩片傳統的瓦楞鐵皮蓋搭而成。

按 1985 年的市價來計算，單是此樑架的建築費用已達到港幣 350 萬元之多。

樑架的設計為一個 5.5×5.5 米的基本格網組合。而主樑的結構深度為 3 米，由 267 毫米直座的鋼通連結而成。當中每 5.5 米的距離為 1.5 米深的橫樑。5.5 米的方格制中，為 0.5 米深的次樑。

燈光的系統裝置，是依結構格網的 5.5 米作基礎編排。每組橫向的燈旁，有 600 毫米闊的鋼架作為燈飾的控制及維修之用。至於球類活動和射籃用的比賽網則排吊在鋼樑架之結構上。

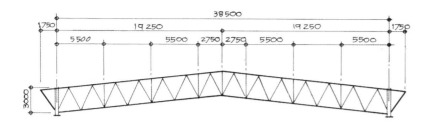

MAIN TRUSS

SECONDARY TRUSS

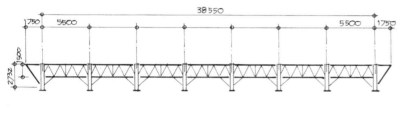

TIE T1

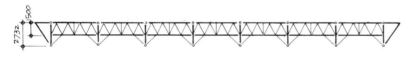

TIE T2

Photo showing the underside of the roof

The structural system of the roof

Interior photo showing the metal roof

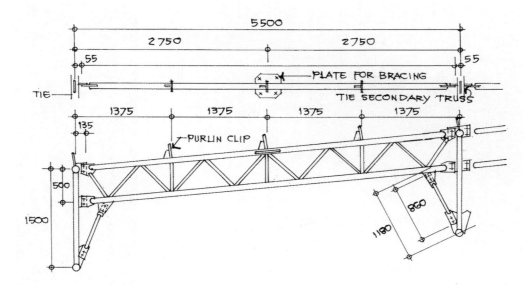

DETAIL OF SECONDARY TRUSS

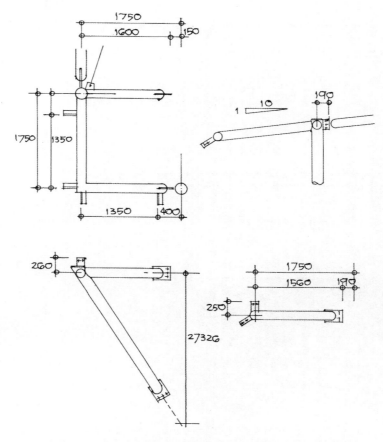

DETAIL OF CORNER TRUSS

Details of different truss systems

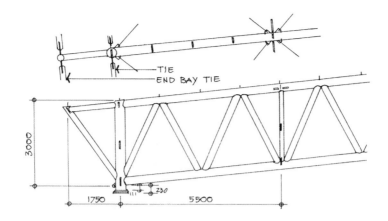

DETAIL OF MAIN TRUSS

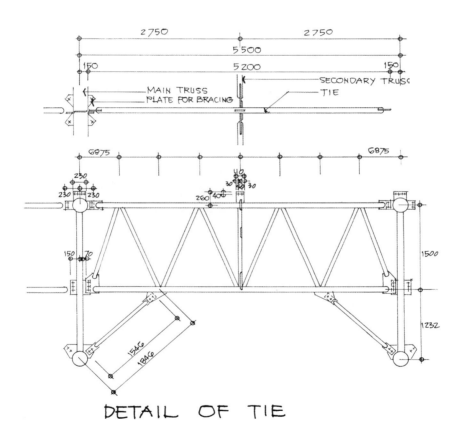

DETAIL OF TIE

Hierachy of the truss

Client:	Urban Services Department
Architect and Engineer:	Architectural Services Department
Main Contractor:	Hung Wan Construction Co. Ltd.
Contractor for Roof Truss:	Okumura Corporation Construction & Engineering Hong Kong Branch
Year of Completion:	1988

3.12

Printing House Vertical Extension: Building on Top of an Existing Building

印刷行大廈：樓層之上的擴建工程

The original building of Printing House was a 13-storey office building between Ice House Street and Duddell Street in the Central District. Due to the revocation of the 'street shadow' regulation in the Building (Planning) Regulations, which previously governed the height of a building in relation to the width of the adjacent street, it is legal to build another eight storeys on top of the existing building, to maximize the development potential of the site.

It is more economical to vertically extend the building than to demolish the existing building and build a totally new structure. In addition to saving construction costs, the existing tenants were not moved out during the construction period, thus maintaining the rental income.

原址中區的雪廠街和都爹利街交界的一幢 13 層高的印刷行大廈，因為受到早期的建築物條例訂定的街影高度限制，發展的總面積不能達到地段和其他的條例能容許的最高標準。然而在這幾年間，因著其他的技術發展，新的標準又再重新考慮。故此，該地段的原有發展，能容許建築物加建八層之多的面積。從利益的角度去看，發展商自然會把握這一個機會。

經計算及考慮後，重建另一座新的大廈比加建另外八層的費用更多。況且在加建的工程進行時，以下的 13 層用戶亦不用遷移，並且要繼續繳付租金。

Structural system

The original structural system was two side cores spanning the office floors in between. Relaxation of the wind code and a modern structural analytical method help to make the vertical structural addition possible. Two more caissons were constructed to add to the original six caissons supporting the larger core. The existing caisson cap was also thickened on the top by 300 mm with reinforcement bars added as shear studs. Holes 4 mm in diameter greater than the bars were drilled from 600 mm to 1,500 mm deep by percussion rotary drill. The holes were then cleaned by compressed air before the steel bars were inserted. Polyester resin grout was then used to fill the holes.

For the superstructure, there were areas (such as modification of the lift shafts) where the existing structure had to be modified and cast with new reinforced concrete. The procedure was to cut the existing concrete without damaging the reinforcement, which was exposed to provide a lap length of 42 times the diameter of the bar. After connection with the new reinforcement had been made, new concrete was cast in place.

Addition of a storey above the existing building

結 構

原來的結構為兩組的核心，承擔著中間樓層的重量。因著科技的進步，新的力學結構計算和新的風力準則，可以容許加建八層的重量。

新地基的結構是在原有的六枝沉箱附近加建另外兩枝沉箱，來承托加建的核心結構。以前的樁帽亦要加厚 300 毫米來承力。以衝擊鑽探在樁帽上，用闊於鋼鐵 4 毫米的鑽嘴，深入至 600 毫米至 1,500 毫米不等，這些孔洞用高壓空氣進行清潔，其後則以聚酯樹脂填補。

至於上蓋結構工程，除了新建的牆身外，還在加建的地方，把原有的混凝土移去，保持鋼鐵結構原好。新加的鋼鐵則需要 42 倍的鋼鐵枝直徑度，澆灌混凝土後便完成。

Delivery of materials and manoeuvring of workmanship

Building materials that arrived at ground level were first loaded onto the sixth floor podium roof. Heavy materials were delivered by machinery, and light materials were transported by existing lifts. A manual pulley system with a steel scaffolding enclosure was set up at the sixth floor podium roof for further transfer of light materials to upper floors. Heavy materials were hoisted by machinery.

External work like the erection of formwork was carried out on a timber plank working platform set up with bamboo scaffolding. The curtain wall and cladding installation were carried out using a gondola system.

Beams with extended steel for later connection

A bamboo catch platform was set up on top of the hoarding for additional safety. Catch platforms were also set up at upper floors together with the scaffolding and green nylon net for protection against any falling objects.

物料的運用與工程施工

建築物料運到大廈的地面層後，重的材料用滑車轉運到六樓的平台上，而比較輕的材料則仍然使用原來的電梯。

鋼鐵架構建於六樓，負責運送物料到達較高的層數。

外牆的工程是依賴木建平台和竹棚架的幫助，而玻璃幕牆的安置則使用吊船。工程進行時，在竹棚之外需加上綠色的尼龍網，防止物件掉到行人路上。

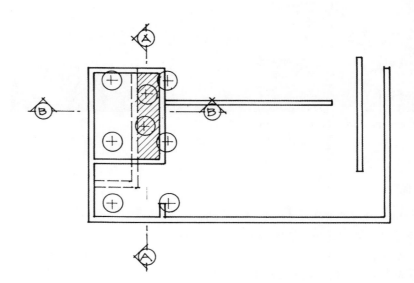

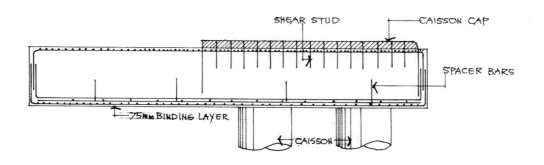

Addition of caissons

Lifts

The three existing lifts were modified to extend eight more floors. The modification was done one by one so that normal operation of the offices could be maintained. Extension of the lift was done only when the new motor room on the new roof was completed. The slabs separating the lift shaft were removed to form the new lift shaft.

電 梯

層數增加後,電梯的數目是不夠應用的,因此,大廈的三部電梯都要調整和改動來應付新的需要。

改動工程並不困難,但電梯的改動則需要待新的電梯槽及機房完成後才可開始。

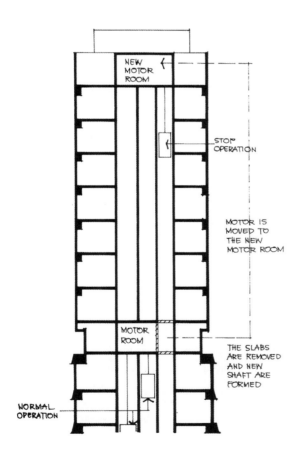

Modification of a lift for extension

Air-conditioning

A larger capacity air-conditioning system was required for the building with the vertical extension. However, the existing air-conditioning units at the 13th floor (the original roof) had to be kept until the vertical extension was completed and the space required by the two-storey headroom was ready. Thus the 14th floor slab was left unbuilt until the new air-conditioning system was installed.

New air-conditioning ductwork and pipes were installed along with the vertical extension. Then the new air-conditioning units were lifted onto the roof by a temporary lifting device from the outside of the building. A Sunday was selected for this operation, requiring partial closure of adjacent Duddell Street.

When the new air-conditioning units had been installed, temporary air ducts connected the new units to the existing ventilation system. These temporary ducts were removed when the new system was completed. Again, a Sunday was chosen to lower the old air conditioners, using lorries with a hydraulic hoist. The 14th floor slab could then be constructed.

The old and new air-conditioning systems are different. The old one was a conventional system with the ductwork in the false ceiling. The new one uses the raised floor system to contain the air-conditioning distribution network.

Worksite with catch platform and hoarding protection

New pipe ducts for services

空調設施

為了供應加建的辦公室的冷氣設備，在 13 樓的原有空調冷氣裝置亦需要增加。由於空調系統的機件龐大，要求有兩層樓底的空間才可以容納、擺放，因此，只有完成兩層樓底的 14 樓之後，才開始放置新容量的冷氣系統。

搬運新的空調系統設施到達機房的時間和方法，都是一個頗傷腦筋的問題。由於中區為一個繁盛熱鬧的地方，車水馬龍，一般的重型機件搬運時間，都選擇在星期日或假期。新的機件由一臨時起重機搬運。至於其他接駁冷氣總風機的風喉和其他小配件，亦由臨時的油壓式起重機運至 14 樓或其他的地方安裝。

在安裝空調系統的概念上，新的與舊的亦有顯著的分別。前者是把所有的喉管收藏在架空地台下，一如香港匯豐銀行總行的設計；而後者就是傳統的收藏在陣底和假天花之上。

In order to maintain normal operation of the offices on existing floors, the extension of liftsfrom the 12th floor to 20th floor was carried out one by one. When one lift was being extended, the other two were still in service. Extension of lifts could be carried out when the extension of all floors above were finished and the space of the lift shaft was provided. Then the motor inthe original lift shaft could be demolished, so the lift could go up to the 20th floor.

After the completion of the lift, it came into service again and the next lift underwent redevelopment.

Water tanks

Relocation of water tanks (potable water, flushing water, and sprinkler) was one of the construction operations that had to be arranged before construction work for the extension part could commence. Four temporary water tanks were installed on the 13th floor (the previous roof), replacing the original water tanks. When the temporary tanks had been inspected and approved for use by the government, the original tanks were then demolished, after which the vertical extension could proceed. At the new roof level, new water tanks were constructed. Simultaneously, the new drinking, flushing, and sprinkler system including downpipes was installed for the vertical extension. After another government inspection and the approval procedure for the new tanks had been satisfied, connections were completed and the temporary tanks on the 13th floor were removed.

水缸

另一項改動設施是水供應系統。這項工程的安裝及編排，需要經政府水務署的批准及同意。

工程分為幾個主要步驟：

1. 由四個臨時水缸代替原來四個放在 13 樓天面的永久水缸。臨時水缸及水喉接妥後，便可開始清拆舊水缸及混凝土隔牆。
2. 同時，開始擴建八層的工程。待新的天面工程完成後，便可遷移新的水缸至適當的位置，然後接駁及安裝新的水喉。
3. 喉管設備完成後需要經政府部門檢驗，滿意後才可以正式使用。

Air-conditioning units on the 13th floor

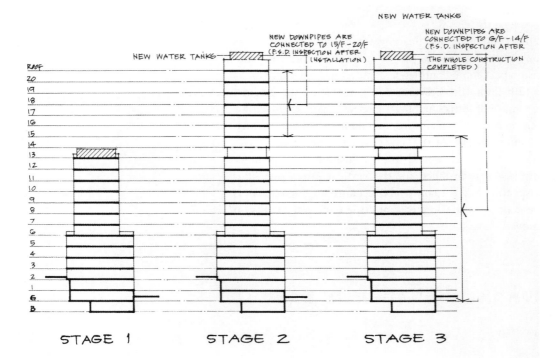

Relocation of the water tanks

Fixing of the cladding

View of the office interior and the curtain wall

Curtain wall and cladding

A curtain wall was used for the external finish of the new vertical extension portion. Cladding with new windows was used for the previous building façade. The core wall is also finished with aluminium cladding without removal of the previous mosaic tiles, to avoid causing possible water leakage points.

The same type of silver reflective glass was used for the whole building. Glass 10 mm thick was used for the typical curtain wall and 8 mm for the replacement of the previous windows. The exposed aluminium surfaces at the curtain wall and the cladding have an anodization of 10.16 micron minimum thickness.

The curtain wall, which is based on a basic module of 1,166 mm high and 1,470 mm wide, offers quite a comfortable vision panel. Every third piece of glazing is openable with heavy-duty hinges to comply with the '1% openable windows' required by government for curtain walls, in case of emergency ventilation. The 1,166 mm module also complies with the 1,100 mm high parapet requirement.

At each floor level separating two floors, mineral wool material with two-hour fire-rating material is installed. The cast-in anchors for basic support of the curtain wall structure have a coat of zinc chromate primer, which is also applied onto in-situ welds.

For the replacement of existing windows, the glass was fixed onto the outside of the window frame by means of structural silicone adhesive, so that from the exterior, the glass appears to be butt-jointed or frameless. Tremco proglaze II (a two-part silicone) was used for this purpose. In conjunction with the silicone sealant, a spacer tape of a silicone compatible urethane double adhesive tape (Norton v-2100) was specified. Also, the cladding itself was set at 55 mm from the surface of the existing finish for the installation.

Photo showing the façade for the old and the new portions of the building

幕牆及外牆裝飾

除了重建的工程外,整幢大廈的外牆亦重新裝飾。主要分為二個部份:

1. 以玻璃幕牆作為新建部份。用 10 毫米厚的銀色反光玻璃,幕牆的模數單位為 1,166 毫米高和 1,470 毫米闊,除符合建物條例規定的 1,100 毫米的欄牆高度要求外,還容許每三個模數為一面可開關的窗,達到「百分之一的總平面面積可開關窗」的緊急通風條件標準。而且,寬敞的窗模數組合可以使內觀環境更舒適。
2. 舊樓的外牆以鋁質板模 10.16 micron 和新的玻璃窗為主。為避免因清除原有瓦磚而導致的漏水情況,新建的鋁板是直接加鋪在外牆之上的,玻璃窗的玻璃仍是幕牆的反光銀色玻璃(8 毫米厚)。鋁板距混凝土外牆為 55 毫米。

在幕牆裝置的條例中,要求層與層之間有兩小時的防火相隔,因此,每一層的連接地方都有纖維綿為防火的物料。

預留在混凝土中的錨,是安裝幕牆的主要結構。在地盤安裝錨的時候,需要加一上層鋅化鉻作保護層,才可以焊接其他構件。

更換舊玻璃的過程也相當複雜。新的玻璃擺放在外面的位置,用結構矽把玻璃接嵌成外牆,亦使玻璃與玻璃之間的接口完全封閉。結構矽為 Tremco proglaze II(是一種由兩部份化學物合成的矽),隔著空間的定位物採用 Norton V-2100,此物質與矽沒有任何的化學反應。

Raised floor

A height of 330 mm is allowed for the construction of a raised floor system which houses the air-conditioning distribution system, connections for electrical supply, and telephone points. The raised floor is made up of a basic modular unit of 600 × 600 × 50 mm thick. The material is a 'sandwich' of an aluminium plate on the top followed by condensed chipboard with cementitious-based backing. A tapered edge affords tightness of installation. The floorboards stand on a galvanized steel frame also of 600 mm^2 grid. Carpet tiles can be fixed on top as the final floor finish.

The whole under-floor space is designed as air outflow and return. Gypsum boards make the space separation. All are covered with fibreglass insulation. The air outlet and return grilles designed to fit the 600 mm grid were installed on the floor with controls and thermostat.

承高的地台

在設計中預留 300 毫米作架空地台,讓中央冷氣系統、電線、電話線等建築設備,可以在地台之下擺放。

地板的基本模數為 600×600×50 毫米。物料為夾心鋁板,在板件底為一層高密度的石板;再用水泥鋪平。地台的結構骨架為 600×600 毫米的鉛水鋼枝。鋁板擺放後,再用 600×600 毫米的地氈塊作最後裝飾。

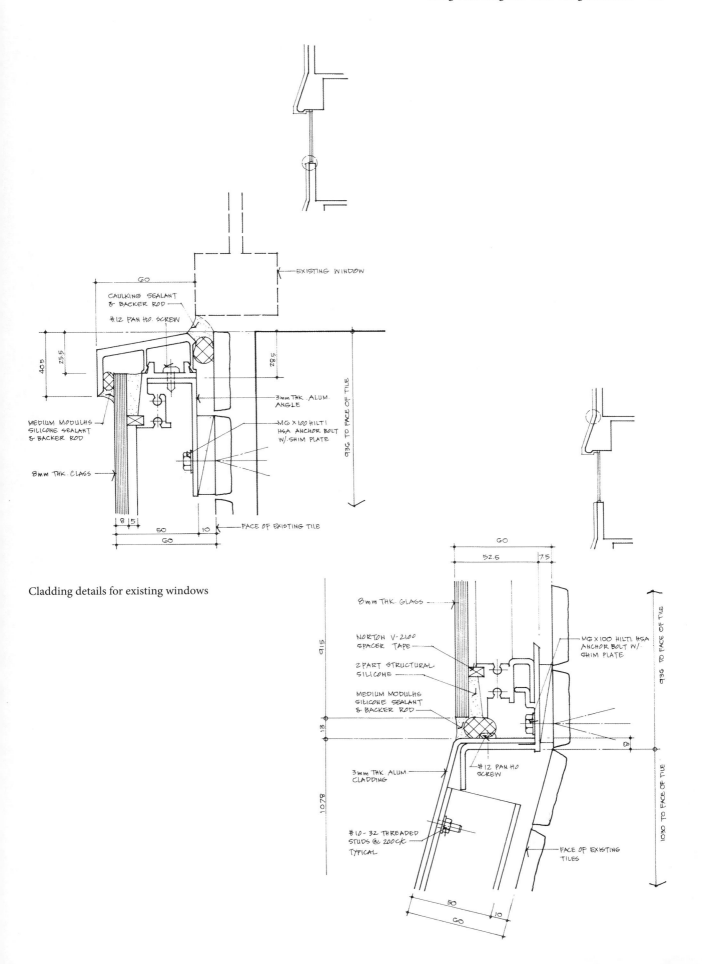

Cladding details for existing windows

Suspended ceiling with integrated lighting

Suspended ceiling

Acoustical tiles form the suspended ceiling, which is suspended by T-section aluminium bearers hung from steel cables to the ribbed slab. Aluminium runners and lighting fixtures rest on the bearers. The space inside the ceiling is for sprinkler pipes and electrical conduits serving the lighting. A recessed type of sprinkler head was also installed for a neat appearance.

假天花

假天花的物質具有吸音功用。每塊標準面積的天花磚由 T 型的鋁質骨架承托著。天花的燈飾設備及空調的百葉，都是依據著天花骨架的模數計劃的。而在假天花之上的空間為消防喉管、電線喉等設施。

Client:	Hang Lung (REA) Ltd.
Architect:	Lee & So and Associates
Engineer:	H. K. Cheng and Associates
Main Contractor:	Hien Lee Engineering Company Ltd.
Curtain Wall and Cladding:	Builders Federal (Hong Kong) Ltd.
Year of Completion:	1991

Raised floor system housing the air-conditioning distribution space

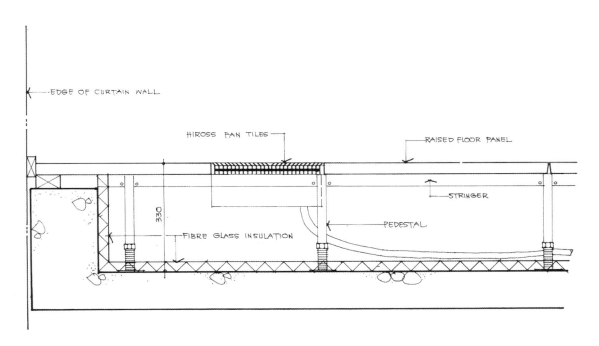

Raised flooring details

3.13
Hong Kong Science Museum: Cavity Wall and External Works
香港科學館：空心牆與外圍工程

The façade of the Hong Kong Science Museum in Tsim Sha Tsui

The Hong Kong Science Museum was designed in a straightforward, simple manner to allow for flexibility and future expansion. It is termed a 'container for exhibits'. Besides the rectangular five-level tower block, a plaza structure was constructed for public use with a series of water displays lit by fibre optics. The total area of the reinforced concrete structure is about 15,000 m^2 and the landscaping area 2.79 hectares. Construction cost was HK$135 million.

香港科學館的設計造型，是一種直接的建築概念及語言，亦具備了未來擴展的機會。由於建築物設計中運用統一的格子作係數，所以它曾被冠以「貨櫃展覽館」的稱號。

博物館樓高五層，建築物周圍乃公眾廣場，總面積為 2.79 公頃，以光纖管的照明品作點綴，有各種不同形式裝飾水池，為晚間本來繁盛的尖沙咀區添上一份優雅。

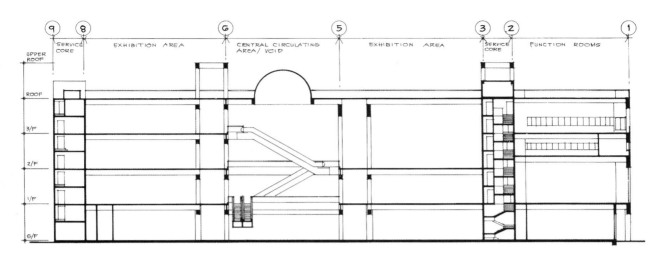

Section of the museum showing the circulation of the space

Grid planning

The planning grid is expressed by different colours on the external wall. The unit length is 4.5 m multiplying and extending to 18 × 18 m. The 4.5 m grid houses all the building services, especially the exposed air-conditioning duct work and lavatories. The 18 m grid contains the exhibition area.

格子設計

設計的基礎源於一組格子。從平面到立體的設計也是依照這個格子，以 4.5 米為所有的建築設施；而 18 米則為主要的展覽場地及空間。顏色的變化與運用亦同時突出了格子係數的設計概念。

Cavity wall construction

To allow for future demolition and extension, the external wall was built with a 4,500 × 4,500 mm panel of cavity wall construction consisting of two layers of 150 mm thick blockwork with a cavity of 50 mm providing insulation and waterproofing. Galvanized steel wall tiles connecting the two layers of blockwork were fixed into the horizontal bed joints during erection and placed at distances apart not exceeding 900 mm horizontally and 450 mm vertically. External tiles of 150 × 150 mm on cement sand rendering form the external finish, and 20 mm gypsum plaster and spray paint form the internal finish.

空心牆的安裝

為了應付未來擴建或修改某部份建築的需要，外牆設計成 4.5 × 4.5 米的空心牆組合。空心牆為兩層 150 毫米厚的磚塊，可以容納防水及隔熱的設備。建造空心牆時，利用鉛水鐵鍋釘固定這兩層的磚牆，當中釘與釘相距應為最多 0.9 米橫向及 0.45 米垂直向。外牆的瓦片為 150 × 150 毫米標準尺寸。以 20 毫米厚的石膏粉作底，待鋪砌好後，才多加一層表面的色油，配合外觀的整體設計。

260 Case Studies

Museum façade expressing the service grid and the exhibition area

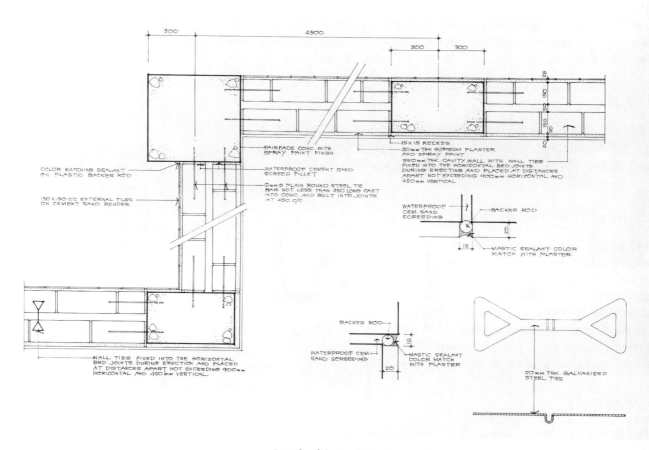

Details of the cavity wall

The conical roof channelling the light to the interior

Fair-faced concrete

Fair-faced concrete was used as columns. Plane end-grained timber was used as formwork, which was struck as early as possible, after which a rubbing-down process by hand using a cork float or block of wood created a smoother surface. Some water was added in the process to form a cement slurry, filling up any air gaps, forming a granular finish. In Hong Kong, the fair-faced finish will weather out with unsightly stains. In the Science Museum, fluorocarbon sprayed paint was added onto the surface to give better protection and to add colour. Also, bolt holes were left in the concrete and expressed as a pattern forming part of the design.

外置混凝土

露天的柱子都是用淨混凝土做成的，故此，木製的板模也要求嚴格。平面的木板塊為注入混凝土時的板模。拆板後的混凝土外牆，再用木板、水松等物來平滑混凝土的表面。在整個過程中，要注入水泥漿來填補空隙，做成粒狀的表面。

然而，在亞熱帶的香港環境中，因為天氣的影響，原混凝土的表面較易受到侵蝕和破損。因此，科學館的原混凝土外牆，要加上一層氟氧碳油作保護膜。同時，這層保護膜亦可為沉灰色的表面帶來其他顏色，以配合其他的外牆裝飾系列。

Conical roof and dome

Though the commonly accepted lighting method for museums is controlled artificial light, the conical roof and the dome do add a bit of natural light to the Science Museum, to realize the architect's idea that light is an important component of life. One hundred twenty-five mm thick concrete forms the structure of these features with plaster and spray paint as finish. Light is transmitted through reflective tempered glass into the interior.

錐形屋頂和圓拱蓋頂

一般博物館或展覽館的光暗設計,都是依賴整套的人工照明方法。但科學館的燈光、照明設計,卻利用天然採光來為單調的空間帶來一點自然的氣氛。

其中的錐形和圓拱蓋的屋頂,把天然光引進展覽廳中,成為這座建築的特色之一。用 125 毫米厚的混凝土為結構,表面噴上一層保護油,光線從反光的強化玻璃射入。

Bridge and walkway

The bridge system surrounding the open plaza affords views at various levels. An interesting pattern of tiles on the plaza floor can be seen. The I-section columns supporting the walkway contain fluorescent light tubes as design features. The balustrades were constructed with aluminium perforated panels and stainless steel angles. The walkway system also leads to a fair-faced concrete tower structure with an interesting staircase design.

橋與欄台

橋的設計讓行人在高處欣賞建築物以外的廣場。I 形切面的結構柱,細意安排原照明設備。還有鋁質的欄杆和不銹鋼的角片細部,把渡橋者引領到小亭之處,都是設計者的一番心思。

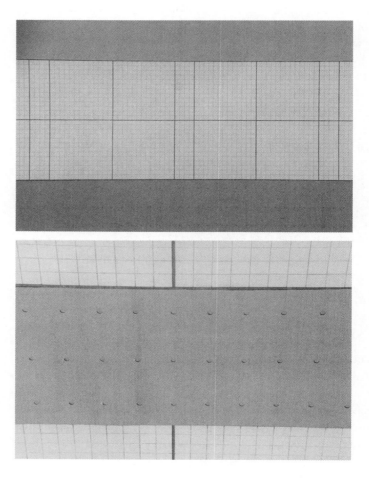

Combination of façade treatment

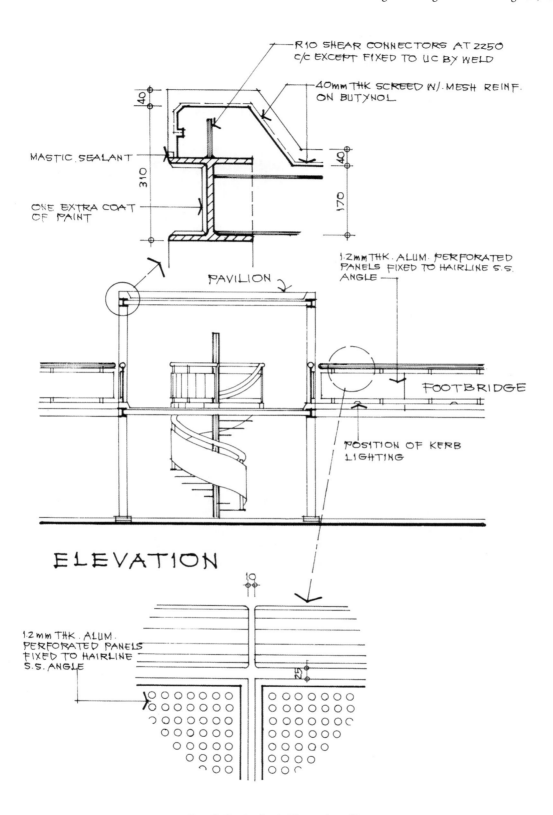

Details for the footbridge and pavilion

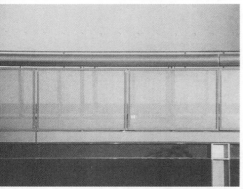

Photos showing the footbridge

Entrance

The entrance is a steel truss structure with polycarbonate cover to form an elegant reflective plane. A mirror stainless steel semi-circular ring projects from the façade. Together with the reflection, a circular ring defines the entrance.

入口樑架

正門的入口是一個鋼樑架的設計。合化碳的外牆裝置做出反光的效果，配合門前的半圓不銹鋼裝飾，剛好成為一個完整的圓形。

Landscaped plaza

A skylight feature is the focus of the plaza. Water seems to originate from here and flow towards a cascade. Light will also penetrate through the skylight into the museum below. Here is a subtle representation of light and water as the 'essence of life'.

On the other end is a semi-circular echo wall. The use of stainless steel contrasts with the artificial granite tiles. Lighting features are incorporated into the wall design.

Steel gateways were constructed for the plaza. This is an elaborate design with stainless steel columns, flanges, brackets, truss, etc. Polished and unpolished surfaces give different colours and textures. Painted colour contrasts with the mechanical bolts and nuts.

廣場花園

玻璃天幕是整個公園的主要地方。花園裏的水流進較低的地方，匯成一個小水池。

玻璃天幕的第二個作用，是使展覽廳獲得天然採光。這樣，水和玻璃天幕構成為廣場能得到「生命之精華」的設計概念。

花園的另一個方位是道半圓形的牆，恰與正門的鏡面不銹鋼裝飾組合成一完整的圓形。

Exposure of services

The air-conditioning ducts are the largest pipes exposed. These ducts come in circular shapes and are made of galvanized strip steel with a locking seam. Spiral pipes are used for fresh air ductwork. The main air condensers are located on the roof. From here, the exposed ducts travel down to the lower floors. A glimpse of these ducts is possible on the external wall as part of the exhibit for the operation of the Science Museum.

建築設備的編排

把一切的建築設備也彰露出來，正好切合科學館的本身意義。

其中最顯眼的要算是鉛水鋼包裝而成的冷氣水喉，主要的冷氣機在建築物的天面，穿來插去的喉管也成為展覽品的一部份，表現科學館的科技運作。

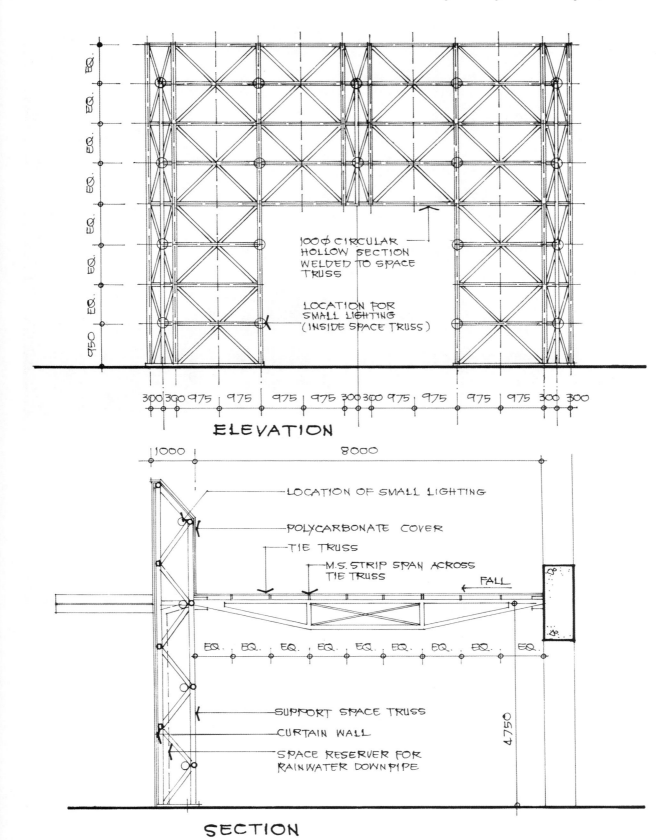

Details of the entrance

Entrance with steel truss structure and polycarbonate cover

Client: Urban Services Department
Architect and Engineer: P & T Architects & Engineers (Hong Kong)
Main Contractor: Leighton Contractors (Asia) Ltd.
Year of Completion: 1990

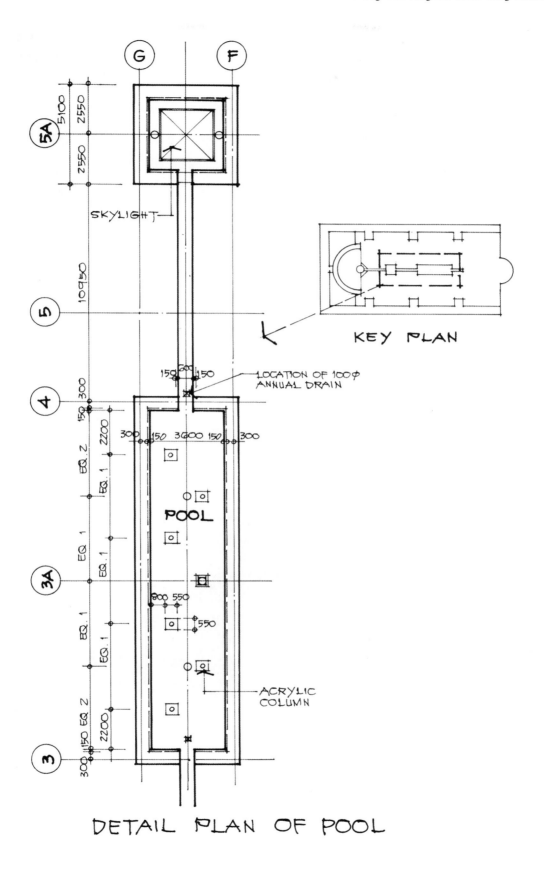

Pool and skylight detail plan

Pool and echo wall

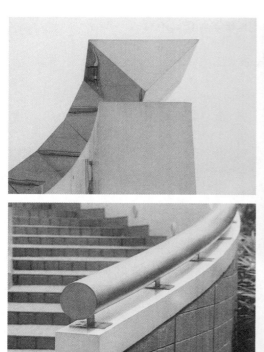

Exposed air-conditioning ducts leading from the roof to the lower floors

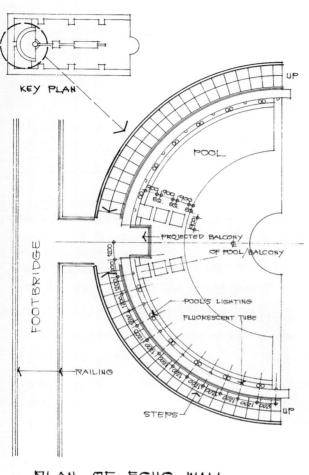
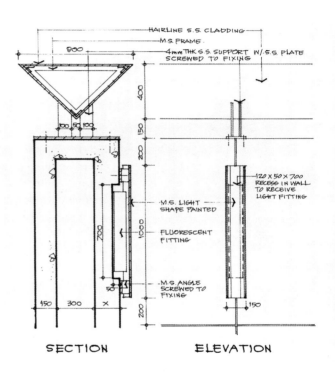
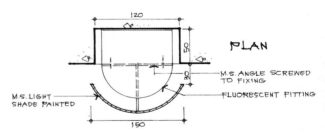

Echo wall details

3.14
Citibank Plaza, Garden Road: An Intelligent Building
萬國寶通廣場：智慧型大廈

Intelligent buildings: The Standard Chartered Bank and HSBC

With the development of building automation, computerization, facilities management, and environmental awareness came 'intelligent' buildings starting in the 1980s. In Hong Kong, buildings with various degrees of 'intelligence' include Exchange Square, the new Standard Chartered Bank headquarters and the HSBC headquarters. At Citibank Plaza, Garden Road (now renamed Three Garden Road, Central) was designed to be a good representation of an intelligent building.

What makes an intelligent building?

Besides the basic function of providing a controlled environment, an intelligent building provides a productive and cost-effective environment through optimization of its four basic elements—structure, systems, services, and management—all interrelated to match the needs of the occupants. Such buildings are designed to accommodate change in a convenient, cost-effective manner and to be capable of providing advanced systems of services and management when needed.

隨著建築物的自動化運作、電腦科技、設施管理及環保意識發展，「智慧型」設計的建築物在八十年代開始便應運而生。在香港，隨著不同程度上的分類，亦有算得上為「智慧型」的設計：中區的香港匯豐銀行總行、渣打銀行大廈、交易廣場，還有在九十年代初花園道上的萬國寶通廣場。

然而，「智慧型」的大廈設計是如何去判別的？

所謂「智慧型」的設計，基本的要求是建築物本身的設計能提供一個「可控制」的環境，而其中又要求這個環境能提供高效率的生產性能，從而達到經濟上的效益。「可控制」的環境主要概括成四個部份：

(1) 結構
(2) 系統
(3) 設施
(4) 管理

這四個層面的設計要求，能符合不同使用者的需要，隨時隨地可作修改、變更，配合新發展的技術和要求，自能成為「智慧型」的建築物。

Building structure

In the case of Citibank Plaza, provisions in the building structure include:

1. a raised floor system to permit quick access to under floor wiring;
2. floor loading of 5 kPa to stand loads imposed by future communication, electronic and other equipment;
3. floor-to-ceiling height of 2.56 m as ample space for various activities;
4. big windows via curtain wall design for vision; and
5. access to utilities with adequate allowance for risers, wiring closets, duct space and so on, and sufficient space in the riser shaft for future additional requirements.

The building structure was also designed to ensure energy efficiency by careful consideration of:

1. building envelope,
2. orientation of the structure itself on site,
3. window treatments,
4. curtain wall or cladding,
5. fixtures, finishes, and furnishings, and
6. thickness of roof and external wall.

These result in energy efficiency and thus reduction in running costs.

Installation of curtain wall (stick system) (top)

建築結構

萬國寶通廣場的建築結構分為：

(1) 提升標準層的地台，使各系統網絡和設施成為最快及最有效率的裝置。
(2) 地台的承重計算係數增至 5 kPa，讓建築物隨著將來的需要，可添置或更改多台的電機、電訊系統。
(3) 地台與天花的平水高達 2.56 米，提供足夠的空間作為日後新系統裝置之用。

Typical floor lobby mock-up (bottom)

(4) 窗的面積和骨架模數亦盡量擴大。
(5) 預留垂直貫通的建築設備管道。

此外，建築結構會同時考慮如何減低能量的耗損，亦把下列的要點計算在內：

(1) 建築物的外觀
(2) 結構的位置設計
(3) 窗的設計處理
(4) 玻璃幕牆或覆蓋板
(5) 飾面與其他設備
(6) 屋頂與外牆用料及厚度

增加能源的效能，同時可節省維修的費用。

Building system

The building system provision incorporating digital data transmission and an optical fibre system can support a high concentration of modern office electronic equipment:

1. Direct digital control is given to heating, ventilation, and air-conditioning, which will optimize the energy efficiency of the whole system.
2. Automatic lighting control is provided for public areas, to minimize energy cost.
3. Back-up generators are provided for emergency power supply.
4. Extra transformers and chillers are for future additional demand.
5. Security card control is for the car park and use of lifts after office hours.
6. Reuter/satellite reception receives current international financial information.

建築系統

由於新的建築系統能應付未來辦公室的需要，所以通訊設備是不可缺少的。運用光纖管及各類新式的電路設計，提供快捷和準確的通訊網絡系統，當然是智慧型建築首要的設計考慮。萬國寶通廣場的系統設計分別如下：

(1) 直接跳字式控制，管理大廈的中央冷氣系統、機電設備。
(2) 全自動燈光控制，可減少電量的耗費。
(3) 後備發電機作緊急時的應用。
(4) 後備變壓器和冷氣機，可應付未來增加的使用量。
(5) 電梯和停車場用電腦化的保安咭。
(6) 安裝衛星天線系統，接收國際通訊的資料。

Basic components of an intelligent building (courtesy of the Great Eagle Company Limited)

274 Case Studies

Raised flooring mock-up showing data, voice, and power transmission

Supended ceiling with integrated lighting fixture

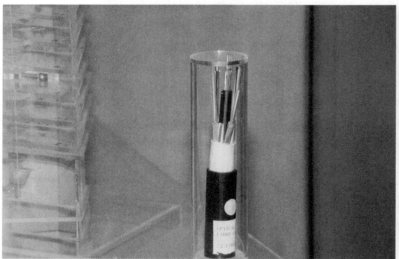

Fibre optics for the building system

Building service study model

Basic components of an intelligent building

Building services

Building services provisions include the following:

1. Voice, data, and video communications to permit integration of demand across the many organizations residing within a single facility, as a basis of the single source service supplier, reducing the overall loading and extending the useful life of the structure
2. Office automation and electronic frameworks for centralized management of communications systems
3. After-hours air-conditioning to small areas is possible
4. Twenty-four-hour air-conditioning to computer/data centres (with back-up power)
5. Efficient and cost-effective accommodation of moves, additions, and changes for tenants' wiring

建築設備

建築設備的分類如下：

(1) 將聲音、資料、影像等不同的通訊系統，簡化為一個統一的通訊網，不但方便應用，更可減低安裝的空間，亦可避免繁複的網絡組合。
(2) 辦公室自動化，提供中央的通訊及控制系統，能更有效率地進行日常商業工作。
(3) 某些地方添加冷氣開關控制，於非使用中央冷氣期間，亦能取得空調。
(4) 電腦中心設有 24 小時的中央空調設備。
(5) 用戶可按著本身的要求和需要，增加或減少設備的接駁。

Building management

For the provision of building management, the 'distributed intelligent design' concept was implemented with innovative distributed microprocessors and computer networking as the main backbone of the system. Management includes maintenance, property, energy and efficiency, security, building services, and directory.

建築管理

智慧大廈的概念是為了方便大廈的管理和組織。全面的電腦化設備，使管理方面更為完善及系統化。

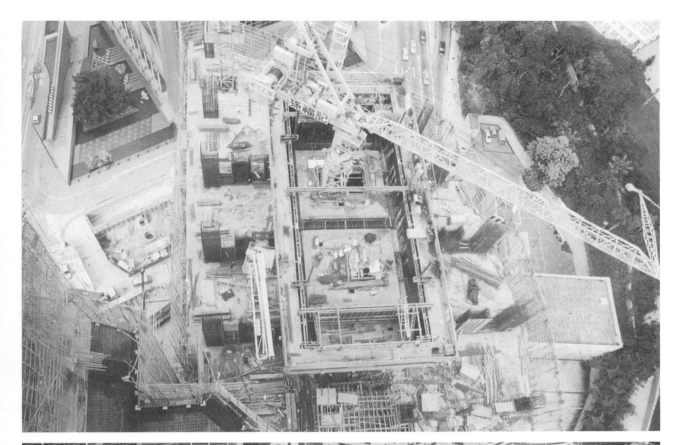

Construction at a three-day cycle per floor with climbform for core and conventional timber formwork for columns and slabs

Climbform for the core construction, working platform, fixing up steelwork, and erecting the inner form after steel fixing

Murray Building, Citibank Plaza, and the Bank of China

Conventional scaffolding Placing boom for concrete Site batching plant

Photo showing the façade of the old wing of the Hong Kong Convention and Exhibition Centre and the surrounding buildings with the newly developed intelligent building in the 1990s

Construction intelligence

The building was constructed with a system formwork with the flexibility to cope with changes in the structural layout and dimensions. Selection of energy-saving reflective glass in the building skin is also a feature in construction intelligence.

The building also achieved a record three-day cycle per floor. For the slab, beams, and columns:

1. The first day was concreting with steel fixing for columns (from 8 a.m. to 3 a.m. the next day).
2. The second day was steel fixing for column and formwork for the slab (from 8 a.m. to 12 midnight).
3. The third day was steel fixing for slabs and beams together with formwork (from 8 a.m. to 10 p.m.).

To cope with this 24-hour concreting cycle, the construction team established a concrete batching plant on site to manufacture all the concrete for the work. This was then pumped up and distributed by a placing boom. Conventional timber formwork was used for the slabs, beams, and columns, but a fibreglass mould was used for ribbed slabs (which reduces the beam depth for longer spans, and thus better ceiling height is created). The construction cycle about two floors ahead of the slab is the core wall, which was constructed by VSL climbform at a three-day cycle per floor: setting up the formwork, fixing reinforcement, and concreting. A shorter time was taken for this cycle, as a smaller work was involved. However, the lift shaft walls were left to be constructed at a later stage.

建築施工的智能技術

實際的建築方法講求效率和完善的計劃及安排。為了應付轉變中的結構設計，承建商計劃一種彈性較高的模板工作。而且選擇外牆反光玻璃方面，亦考慮到玻璃的反熱性能和反光程度，減低因熱能侵入室內的環境而增加能量的消耗。

建築的速率以三天完成一層為紀錄：

(1) 第一天，由早上八時至凌晨的三時，完成柱的紮鐵及澆灌混凝土工作；
(2) 第二天，由早上八時至深夜十二時，地台的紮鐵及柱的模板工程；
(3) 第三天，由早上八時至夜間十時，完成地台層及樑的模板工作。

而為了保持 24 小時的澆灌混凝土工程和更有效率的安排，承建商在地盤中設置一個混凝土的拌合機，把混凝土分發到所需要的地方去。

地台層、柱和樑都使用傳統的木板模和紮鐵工作，但肋板則以纖維模來簡化建造程序。而肋板的優點是可避免因為樑的深度過深而減低樓身的高度。

建造週期在兩層之上是核心牆，核心牆的工程以 VSL Climbform 滑模方法，由紮鐵、釘板到澆灌混凝土也是維持在三天之內完成。餘下的升降機槽則安排在進度表的另一個工程時間。

Summary of additional duct spaces and plant room space requirements for various computer and telecommunication systems which are required to fulfil an intelligent building design concept (courtesy of the Great Eagle Company Limited)
一般設備空間的尺寸

	Duct space	Plant room space
Central digital	1,500 mm (W) × 900 mm (D)	
Office automation and central database computer	200 mm (W) × 50 mm (D)	650 mm
Computer backbone system	1,000 mm (W) × 900 mm (D)	
International financial information system	1,200 mm (W) × 900 mm (D)	A space on the roof for accommodating a 1.4 m diameter satellite disc
Broadband coaxial cable system for video signal transmission network	400 mm (W) × 9,000 mm (D)	Nil
Video conference room	Nil	80 m^2

If all the systems are installed, a common duct shaft 3,100 mm (W) × 900 mm (D) should be reserved.

Developer:	Shine Hill Development Limited
Developer's Agent:	The Great Eagle Estate Agents Limited
Client:	SFK Construction Management Limited
Architect:	Rocco Design Partners
Structural Engineer:	Ove Arup & Partners Hong Kong Limited
Mechanical and Electrical Engineer:	J. Roger Preston & Partners
Quantity Surveyor:	Levett & Bailey
Contractor:	SFK Construction Management Limited
Completed:	1992

3.15
Central Plaza: Super High-Rise Concreting Technology
中環廣場：超級高廈的混凝土建築技術

To justify the high cost of land and the prime location, the architect's main duty was to design a prominent commercial building which will stand out from its neighbours and enjoy the best harbour views. A building of 368 m high which was the tallest in Hong Kong (at the time of completion) provided the design solution. The triangular plan form gives the best orientation for views and generates a dynamic grid for detail design.

From the bottom, the building consists of three levels of basement for parking cars. The ground floor is mainly for pedestrians and has a landscaped area. The first floor is a bridge system connected to the adjacent buildings. The second floor, at 13 m above ground, is the 12 m high office lobby. Then a transfer structure supports the 70 levels of floor slabs above. The 50 m high pinnacle on top crowns the tower.

Technically, the height of the tower and the high speed required for construction make this study an interesting one. Furthermore, this represents one of the most advanced concrete technologies for construction.

地段的位置往往是決定土地價值的主要因素。面對維多利亞港，「中環廣場」的位置可謂名正言順的灣仔「地王」了。而在此高昂地價的地段上，建築師如何著手設計，使建築物成為那裏的標誌，也是值得我們研究的。

中環廣場是一幢樓高 368 米的商業大廈，也是目前香港最高的建築物。從三角的建築物平面圖看來，除了是動感之外，在設計概念中亦象徵著三個不同方向的行人天橋的連接網絡。至於建築物的編排方面：地面之下為三層公眾停車場，地面層連接一個市政局的公園為公眾綠化區及休憩地方；而 12 米高的辦公室大堂設在距地面 13 米高的平台上。大堂上的混凝土轉力層，承托著 70 層的樓宇結構和 50 米高的屋頂及裝飾。

畢挺的高樓與它高速的施工時間，還得配合先進的混凝土建造技術。

Contractor's site organization

The site was operated by a joint venture management team from the developers, who also formed a joint venture in-house construction company, Manloze Limited. The management team reported to the developers and transferred instructions from various parties to the construction company as well as controlled payment. This system was a highly efficient operation.

The in-house construction company mainly worked on cost-saving, as the contractor's profit can be reduced by contractual arrangement, which is a management type with prime costs plus zero profit. To speed up construction, detail design could be worked out with the contractor to ensure economical and easy-to-build solutions. There was also no claim for both cost and extension of time, which is a tedious job for consultants.

There were about 1,000 workers on site during the construction of a typical floor. Four passenger hoists and four cargo hoists were installed on the outside face of the tower to facilitate vertical transportation. Three concrete pumps were set up for transportation of wet concrete to the required floor. Moreover, three tower cranes were mounted on the building to reach every corner of the site. All these factors contributed to greater efficiency.

Compared to the huge volume (173,000 m^2 office space) of the building, the 7,230 m^2 site is not large. Thus there was little room for storage of materials. The ground level was mainly reserved for incoming vehicles delivering concrete and other building materials and some steel reinforcement work.

Taking the above measures for efficiency, a typical floor can go up in four days per storey and the building completed in several phases to get the best income for the developers.

承建商的地盤計劃

負責地盤工程的承建商為發展公司的內部公司。因此，從指示、分派員工、訂料及工資等事項，都有完善的安排及監管。而慣常在施工期間出現的問題和有關合同的糾紛及爭議，都可以同時避免。有時為配合施工的進度，承建商會特別注意細部的設計及主動地考慮實際的施工步驟。這樣可以減少工程被延誤的機會，同時亦可以間接地控制建築的一切費用；對發展商來說，更能達到時間和金錢上的全面管理。

施工期間，單是建築標準層時便約有 1,000 名工人；兼有四輛客用及貨用電梯掛在建築物的外圍牆。三組的混凝土泵和起重機，負責三角形平面的每一邊，使施工步驟更有系統、快捷和簡單。

總建築面積為 173,000 平方米，與地盤的 7,230 平方米比較實在相差很遠。地盤以地面層作擺放建築物料用。地盤的工作進度曾以四天之內完成一層為紀錄。

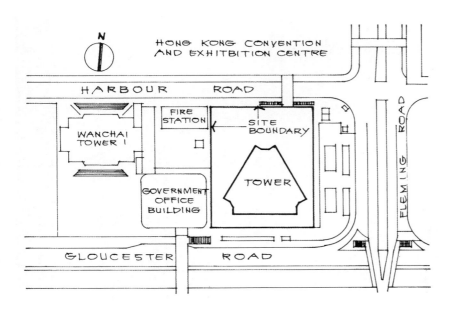

Master plan of Central Plaza

Bamboo scaffolding and steel propping used for the tower base and installation of granite slabs. Upper floors were constructed by system formwork. The curtain wall was installed by a unitized system.

Diaphragm wall and caissons

A 800 mm thick diaphragm wall was the first structure to be constructed on the site. This was built along the perimeter of the whole site within six months. The procedure was to employ a chisel-like machine to dig the wall. Then liquid bentonite was pumped to replace the soil. Finally, reinforcement was put in, and pumped concrete was used to displace the bentonite. The depth was about 35 m and was grouted onto solid rock to create a watertight box for digging the caissons. This also allowed the superstructure to go up once the caissons had been constructed and the basement to be built by the top-down method. Moreover, the diaphragm wall can distribute some of the wind shear.

Caissons are designed to sit on solid rock with bell-outs and are about 30 m deep. The maximum size of caissons constructed on this site was 7.8 m in diameter. The large size enabled the machine-dug rather than the hand-dug method. To save extra time spent for excavation, the pile cap was designed at the second basement level rather than conventionally below the third (lowest) basement level.

Small forklift trunk transporting materials within the site

地下連續牆和沉箱

環繞地盤周圍的 800 毫米闊地下連續牆，是首項的結構工程，共花了半年的時間完成。先用挖泥機挖掘，把液體狀態的皂土（bentonite）注入泥裏，由於皂土的密度比泥為重，泥土便因此而被榨壓出來。接著紮鐵和灌注混凝土入牆中，皂土的密度亦較混凝土為輕，反被榨壓至地面上。35 米深的地下連續牆的底部與地下石相接，再以防水漿填補之間的隙，使地庫層保持乾爽。

承建商採用「由上至下」的地庫施建方法，使上層結構的建築工程可以同時進行，況且，已先建築好的地下連續牆具有承力牆的作用，可抵擋風力。

30 米深的沉箱結構設計，較一般常用的直徑略大（最大的直徑為 7.8 米），而且在沉箱接連地下石的部份擴闊，這樣的設計可以容許用機器作為挖掘，避免人手操作的危險和減低施工的時間。至於樁帽的設計，亦由一般最低層提升至第二層的地庫台階之下，可減少挖掘的時間和困難。

Trucks delivering concrete and steel fixing on site

Top-down method for basement construction

The top-down method enabled simultaneous construction of the basement floors and the superstructure after the caissons had been completed.

Supported by structural steel inside the caissons, the ground floor slab covering the whole site was constructed first. A large opening was allowed in the ground floor slab to allow for excavation of the basement, which was one floor after another and from top to bottom. Concrete was then placed around the previous structural steel to form the basement columns. At the same time as all the top-down work was done, the structural core of the building was built from the ground floor upwards.

Temporary lifts transporting workers and gondola for curtain wall fixing

由上而下的地庫建築方法

先前曾提及的「由上而下」的地庫建築方法，是由於沉箱內的結構鋼作了承重的保護，地面的台階能首先做好。在地面的台階預留一個缺口，作為地庫的其他台階的挖掘和建築。完成後，才把混凝土注入柱模成為永久的結構；與此同時，地面之上的結構亦可一併進行。

Construction of tower base

The tower base columns are 2.6 m in diameter and 25 m high. Steel column formwork (Paschal Trapezoidal Girder Formwork System) was used for casting these columns. They were designed to climb to their required height with the use of climbers attached to the form together with a special anchor.

These columns support the 5.5 m deep transfer structure, which allows the typical floor columns to be deleted alternatively. Traditional timber formwork was used, and steel propping was used as a temporary support for the heavy transfer structure.

建築物的塔底建造

建築物的著地柱子為直徑 2.6 米、高 25 米、承著 5 米深的轉力層，藉以減少標準層的結構柱。承建商以鋼鐵的柱模來建築這些巨型柱子，鋼鐵模板設有爬梯和特別的錨作穩定柱身，亦可以伸縮到適當的高度。而轉力層的模板則沿用傳統的木板模件。

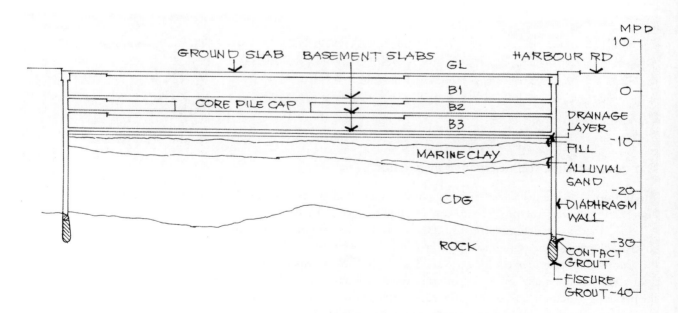

Section showing the diaphragm wall

Grade 60 concrete

Grade 60 concrete acquiring a 28-day cube strength of 60 N per mm^2 was used in order to reduce the size of the structural columns, thus gaining more usable floor area. This was the first time a private-sector project in Hong Kong used such concrete. Considerable research and tests were carried out using a trial mix and mock-ups of the large-diameter columns to study the temperature effects during curing of the concrete. Fuel ash replacing cement can reduce the hydration heat evolved.

As a result, a water cooling system was installed before concreting. This involved the installation of water pipes in the columns before concreting. After concreting, the continuous flow of water was maintained at least 36 hours to remove the excessive heat. This system was used for reinforced concrete columns in the basement, podium, and tower. Grade 60 concrete was found to be more economical than is ordinary concrete mix, which requires larger structural member sizes.

To keep strict quality control, the contractor had quality checking done by the concrete factory on the site. Resident engineers tested the concrete samples randomly and frequently.

Steel column at the basement cladded with concrete

60型混凝土

中環廣場運用的 60 型混凝土在香港的建築業中仍是罕見。它只需要 28 天的時間，便能使混凝土的承力點達到 60 N/mm^2。因此，採用 60 型混凝土，是要使結構柱的面積盡量減少，而相對增加標準層的實用面積。

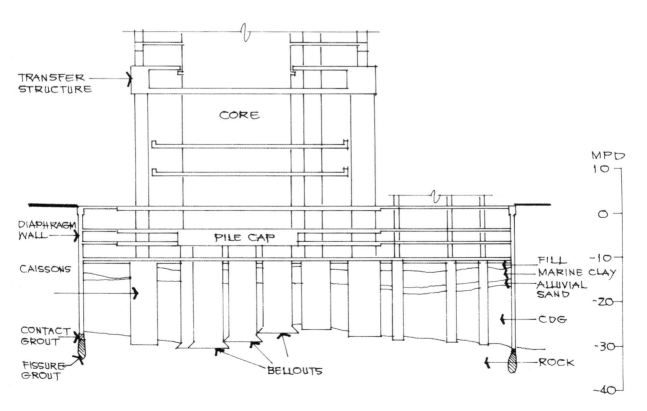

Foundation, basement, and tower base

但由於 60 型混凝土是本地私人發展商首次採用，承建商需要呈交混凝土的測試及檢驗報告給政府批核。模擬測試還包括量度混凝土凝固後的溫度。所以，為了進一步降低及控制凝固期間所散發的熱量，還以燃料灰代替一般慣用的水泥作混合料。

澆灌混凝土前，在柱的模板之內擺放冷水管；澆灌完成後，冷水不斷流注入柱的管導內。經過最少 36 小時不斷的冷卻作用，混凝土可以達到預期的效果。從經濟的角度計算，採用 60 型混凝土作大型的結構，較用普通混凝土更為便宜。

品質監管方面，除了承建商聘請的測試混凝土公司長駐地盤作為專業顧問外，還會有工程師在地盤定期抽樣檢查品質。

Delivery of concrete

Due to the height of the building, delivery of concrete could not be by the traditional method, so a pumping system was devised for the vertical transportation of concrete up to a record height of 308 m.

Two sets of operating pumps and one set of standby pumps were located near the site entrance, to ensure the smooth flow of concrete-delivering trucks. Pumps were changed to different pressures to serve different zones of the building height.

These sets of pipelines were installed in a lift shaft to serve from the ground floor to the concreting area and were changed entirely after 25 floors of usage.

Actual concreting was by the three sets of placing booms located to cover the whole floor. Floor-climbing devices enabled the placing booms to rise to the upper floors after casting of the floor slabs.

混凝土的運送

由於建築物的高度，所以一般運載混凝土的方法會比較困難。為解決運送的難題，於是以電動泵把混凝土運至高於地面 308 米的位置去。

兩組的電動泵在一般情況下運作，另一組則作後備。為方便運送混凝土的車子到達現場，所有的泵管都安排放置在地盤的入口附近。泵的壓力也隨著建築物的高度而調節。澆灌混凝土時，在該層中擺放三組灌注機，使混凝土能均勻分佈。

另一方面，運輸的管道經由升降機槽內進出，在完成 25 層後，所有的管道都要全新更換，方能保持劃一的水準。

Climbform for the core wall

Being the first structure to go up in the atypical floor, the VSL climbform system as a self-lifting method served the core wall formwork. It is made up of an outside trussed frame as the external platform, internal platforms with two working levels, plus hydraulics for the 'climbing' action. Also a tolerance range of ± 10 mm per 4.0 metre height was achieved. To cope with the triangular configuration of the core wall, three sets of independent climbforms were installed.

A three-day cycle was achievable for each set of the climbform. The external platform was raised by extending the climbform pistons pushing off the completed concrete walls. The climbing operation took about two hours. Preformed seating sockets were used to insert the platform support pins. The

VSL climbform for the core wall

Steel couplers for the extension of reinforcement

Steel formwork for the tower base columns

Steel propping for the transfer plate

Timber formwork for the transfer plate

A sample column for testing

Placing boom for concrete

Three sets of pipelines delivering concrete to the upper floors

Concrete delivery to the site

external forms were then cleaned and oiled with the platform levelled and plumbed. Reinforcement was then fixed, after which internal platforms were raised by means of rodjacks or long-stroke jacks. The forms were then closed and tied ready for the next pour. After the concrete pour was finished, form ties were removed and all forms stripped. Internal forms were then cleaned and oiled.

核心工作平台

因為發展商希望施工期盡量縮減,所以嘗試採用一種較簡單和容易操作的輔助工具——VSL 升降台,以進行高層的釘板及紮鐵工作。

VSL 升降台主要是為核心牆的板模而運作。配合三角形的平面設計,在三角形的每一邊都設一獨立的鋼鐵架,外圍架的內層為一個以油壓式操作及可達兩層高的工作台,當升降台提升至 4 米的高度時,其偏差程度可控制至 ±10 毫米。

每組的升降台可以在三天之內完成三分之一層的核心牆。完成一個部份的工作後,升降台可由外圍已完成的混凝土牆推上,每次的程序花費兩個小時。

Rectangular perimeter columns

The IFS panels with Paschal key bolts for connections were used for the perimeter columns. Walkway brackets were incorporated to provide safety to workers without scaffolding. Also, no tie bars were required, giving a high-quality finish. Reductions in column size could also be accommodated. To avoid oversizing steel at the overlap joints, couplers were used to join reinforcement bars.

方形柱模板

方形柱的建造用 IFS 板模及 Paschal 的鍵螺栓的模件,其優點是它本身已附有工作用的小平台,可以免卻一般以竹棚為工作台的危險性;而且,由於鍵螺栓已經有拉桿的作用,可以提高完成後的混凝土柱的質素;同時,更可避免混凝土柱因板模的厚度而縮減。在建造混凝土柱時,為了減少結構所用鋼鐵的數量超過預算,改由耦合器來連接鋼鐵,亦是很好的方法。

Table form

For concreting of the atypical floor slab, the table form system based on the use of small tables was employed. Each structural bay was covered by four to six small tables, also coping with adjustments in design layout.

After removal of the column forms, the edge tables were hoisted in place. The edge tables could allow installation of the curtain wall support brackets, which in turn catered for reduction in perimeter beam sizes. Then the internal tables were hoisted in place. Plywood for the beam soffits were also positioned. Then the tables were fixed by tie bars. Vertical props were placed under the beam soffits before concreting.

The four-day cycle was achieved. The first day saw the completion of the column, the second day the table formwork, the third day the reinforcement, and the fourth day dealt with the pouring of concrete. Two sets of internal tables and one set of edge tables were provided, because the latter could be removed within 36 hours due to the shorter span of the edge beams.

枱模

以枱模方法來建標準層的地台，配合不同的設計單位，分別由四至六個小枱模完成。

柱子完成後，旁邊的枱模則提升到適當的位置。每塊枱模都自行預先裝好玻璃幕牆的結構鈎。圍邊的枱模安妥後，內枱模可以擺放在正確的位置，再由拉桿固定它們結構樑的模板，而支撐工作亦同時進行。

以枱模工作作標準層的施工，可以在四天內完成一層單位。首天完成了柱的工作，第二天開始枱模工作，第三天紮鐵，最後為澆灌混凝土，整項工程共運用兩組的內部枱模及一組旁邊的枱模。旁邊枱模可在36小時內拆掉作另一層的工作。

The external platform was raised by extending the climbform pistons and pushing off the concrete walls of the previous pour. During the operation, external and corridor forms were cleaned and oiled. After lifting, the platform support pins were inserted into preformed seating pockets. The platform was then levelled and plumbed after reinforcement fixing and carpentry work was complete. The internal platforms were raised by means of rodjacks or long-stroke jacks. The forms were then closed and tied in readiness for the next pour, when the completed form ties were removed and all forms stripped. Internal forms may be cleaned and oiled at this stage.

Slab formwork

Column formwork

Construction of the rectangular perimeter columns

Curtain wall

The unitized system of curtain wall was used to give a fast installation method. The prefabricated panels were stored on each floor. A hoist was used to erect the curtain wall units. A gondola was used to carry the workers for the installation work.

玻璃幕牆

組合式玻璃幕牆的方法,是以預製的模件先在廠中加工嵌接,之後再在現場安裝。現場除了作儲放預製的模件外,還備有吊車作安裝模件之用。

Phased completion

To give an early return for the clients' investment, three phases of completion were planned for the building. The first phase was up to the 27th floor, the second phase the 45th floor, and the final phase covered the remaining floors above the sky lobby.

Lift zones and building services were arranged to match each phase of completion with minimum temporary connection work, just like three independent medium-rise buildings stacked on top of each other.

完成階段

為配合發展商的投資計劃,大廈分為三個階段完成。首階段至二十七樓;第二階段為二十八至四十五樓;最後一個階段為四十六至高層的觀景大堂。

升降機的設計施工也得同時配合三個階段的工程,以減少不必要的臨時工序,情況就類似三幢獨立的中型高度樓宇加疊聯合起來。

External platform system

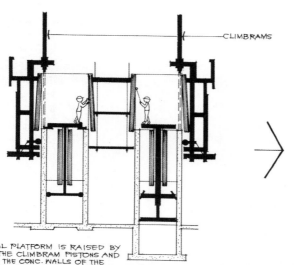

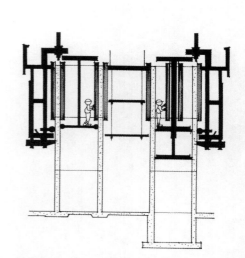

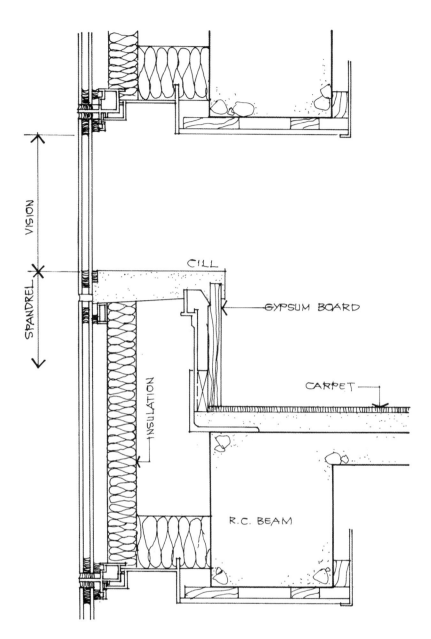

Typical details of a curtain wall

Gondola for curtain wall fixing

Unitized system for curtain wall

294 Case Studies

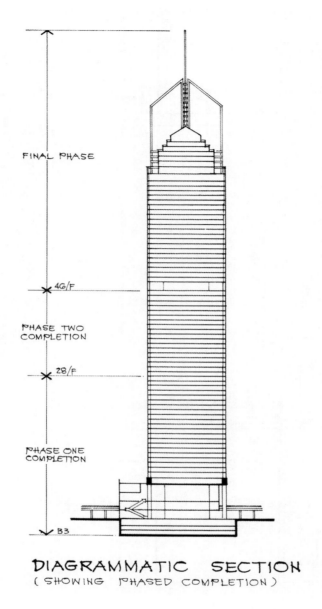

Diagrammatic section showing the phased completion of the tower

Client:	Sino Land Co. Ltd., Sun Hung Kai Property Development Co. Ltd., Ryoden Property Development Co. Ltd.
Architect:	Ng Chun Man & Associates Architects & Engineers (HK) Ltd.
Structural Engineer:	Ove Arup and Partners
Electrical and Mechanical Consultant:	Associated Consulting Engineers
Quantity Surveyor:	Levett and Bailey Chartered Surveyors
Main Contractor:	Manloze Limited
Year of Completion:	1992

3.16 Sam Tung Uk Museum: Landscaping and External Works
三棟屋博物館：園林設計

Photo showing the cluster of rooftops at Sam Tung Uk

Built about 200 years ago by the Chan family, and now turned into a folk museum, Sam Tung Uk, as the name means, contains three rows of houses, the fourth added at the rear and at a later stage. This is a good example of a single-clan walled village.

今日的三棟屋博物館，前身原是二百年前一位陳姓的祖屋。原址建築群為三行並排而列的房舍，乃取其設計的組合，故以「三棟屋」命名。現經政府重新修建，在三棟之外還多建第四棟並排的房舍。三棟屋可說是傳統圍村式的民居建築。

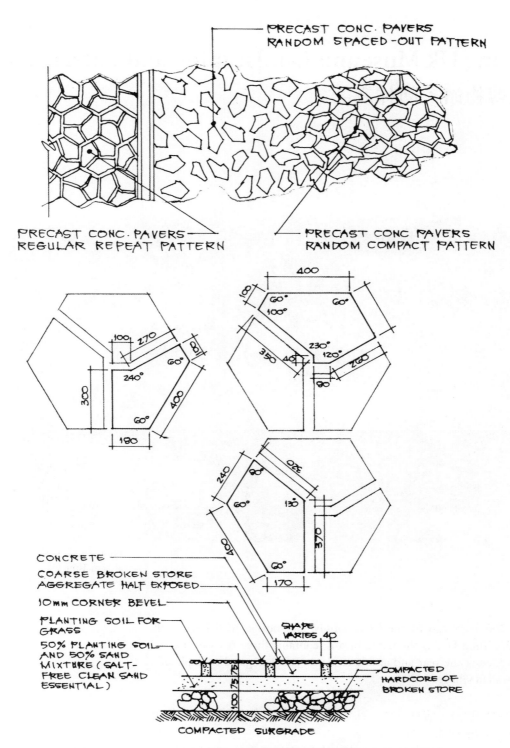

The precast concrete pacers in the landscaped area

Landscaping concept

A natural and somewhat primitive setting was the concept for this 6,000 m² garden. Terraces simulating the cultivation of fields in the countryside slopes were used as the main landscaping element and as a backdrop to the museum. Folk culture is experienced in the museum as well as in the surrounding landscape. Even the pond was designed with a rustic appearance on the edges. This is the case when irregularities and non-alignment are arranged with subtle details to form controlled vistas. The line of vision along the meandering paths is an important aspect of the design.

園林設計的概念

博物館坐落的地方，有著一個偌大的後花園，而這 6,000 平方米的後園設計，充滿著自然和諧的格調。高低起伏的台階，穿插著細意安排的綠草和花卉，有如傳統鄉間所見的梯田樹苗，更顯出博物館建築物的特色。一般中國園林的意境，諸如亭台流水，迂迴曲折的小徑，在此處自是盡見一斑。

Precast concrete pavers

The artificial terraces in a farm-like setting

Case Studies

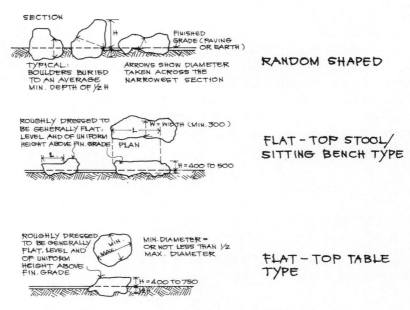

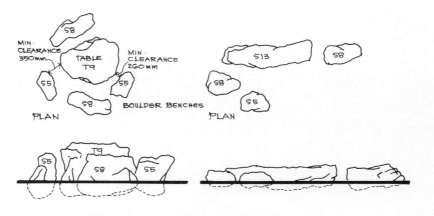

Design of a sitting group with boulders

Photos showing the arrangement of boulders as a sitting area

Landscape details

To form the terraces, a 300 mm wide stone wall about 500 mm high was built as a toe wall. Subsoil drains with stone surround were installed to avoid flooding. A reinforced concrete footing was constructed as the base.

The precast concrete slabs finished with exposed aggregate for paving are designed to form either a regular hexagonal pattern or a random pattern. The slabs were then set in 75 mm thick, 50% sand 50% soil mix and on top of a 100 mm thick hard core.

Granite boulders are natural building material compatible with the landscape. Some are naturally finished and set to hide and contrast with artificial signboards and parapets. Others are cut flat and used as tables and seats. The placing of these boulders is usually done onsite rather than on the drawing board.

Bentonite lining of 200 mm thick forms the base of the pond. This was selected for its workability over irregular surfaces. A boulder layer of 200 mm thick forms the finishing base. An artificial stream of a 200 mm reinforced concrete base and waterproof cement sand rendering leads water into the pond. Boulders were added along the stream for a natural look.

園林組織的細部

起伏的台階以 300 毫米闊及 500 毫米高的石牆組成。為了防止積水，大石堆的周圍設有地下疏水管，而底層則為結構混凝土。

預製的混凝土層面之上以粒料構成六角形或不規則的圖案，最後用 75 毫米粗的混合料（沙和石各佔半份）附在 100 毫米厚的硬層之上。

園中的天然麻石，形態和大小各異，都是依據設計者的構思，放置在現場環境的擺設，除了作為造境之用，更是遮蔽或對比園內的牌碑和公用設施。

環繞荷花池的旁邊為 200 毫米厚的皂土，其特性可隨著水池的邊圍，作不規則的線條組合。而在水池的另一邊為一人工的小流川，以防水泥沙在 200 毫米厚的混凝土層之上建成。周圍還放置不同形態的天然石堆群以作點綴。

The pathway and the pond

Gatehouse

The gatehouse represents a modern way to have a traditional Chinese building. The roof is constructed of 100 mm thick structural concrete slabs. An asphalt waterproofing layer was applied before the Chinese clay roof tiles were laid with mortar and wire mesh reinforcement. Underneath the concrete slab is another layer of clay tiles supported by wooden battens and 150 mm diameter timber purlins which are fixed with ends onto the concrete wall.

Natural granite architrave and a threshold with hinge block and granite socket block were used to fix the wooden door, which is also constructed in traditional style with a wooden latch lock.

Flooring is 30 mm exposed aggregate concrete paving on reinforced concrete slab set on waterproof membrane, blinding layer, hard core, and compacted subgrade.

The gatehouse

The final product of the gatehouse with its storeroom and kiosk harmonizes with the surrounding landscaping and merges with the original Sam Tung Uk Museum. Much of the success is due to the concern with details and correct use of building materials and techniques.

門房

門房在傳統的中國建築中，有著特別的意義。新建的門房以新的建築模式把它的象徵重現在這博物館中。

屋頂變為 100 毫米厚的結構混凝土，在防水瀝青層之上才鋪放中國式的天面黏土瓦。屋頂的底部亦鋪放同一的黏土瓦，另更仿效古法用木結構而加上木條和 150 毫米直徑的桁條。

天然麻石作為門頭和門腳，配合木構成的門栓，都是仿照傳統中國的木門樣子而設計的。

地面亦由粒料構成不規則的圖案，使鄉村味道更濃。仿古細部設計，使新的門房為博物館增添幾份懷舊氣氛。

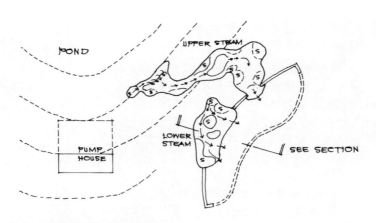

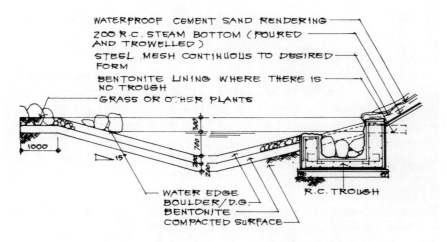

Plan of the pond and its profile

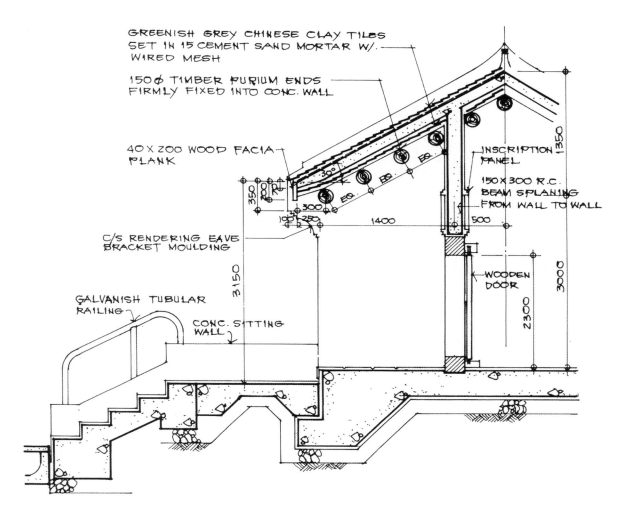

Section of the gatehouse

Chinese clay tile roof

Timber purlins

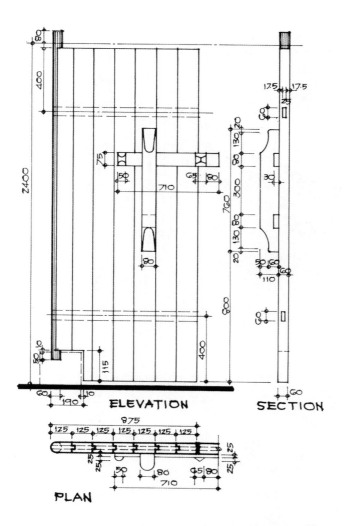

Wooden door of traditional construction

GATEHOUSE DOOR CONSTRUCTION DETAILS

Gatehouse door construction details

Client:	Tsuen Wan New Town Development Office, New Territories Development Department
Architect:	Patrick S. S. Lau of Design Consultants
Landscape Architect:	C. K. Wong
Contractor:	Yan Lee Construction Co. Ltd.
Year of Completion:	1988

Chapter 4

Drawing Practices: From Design Sketches to Tender Drawings

從草圖到招標圖

Kee Yee Chun, Tris 祁宜臻

> I prefer drawing to talking. Drawing is faster, and leaves less room for lies.
>
> — Le Corbusier

Drawings are the fundamental tools for architects to communicate. This communication exists on multiple hierarchical levels, ranging from that between architects and clients, architects and contractors, to contractors and suppliers. At the very gestation stage of a design concept, the sketch is often the architect's first tool to explore different ideas. This earliest stage of concept formation is bounced back and forth, traced and retraced on the sketch papers—a fundamental way to communicate between the eye and the mind. When the architect begins to move from the initial conceptual stage to subsequent stages of built work, these drawings take on a completely different set of tones and voices. For instance, at the conception stage, drawings tend to be less technical and more poetic and impressionistic. The objective of these first drawings is to provide the client the opportunity to endorse a design direction from a decidedly emotional level.

> Drawings are the expressions of one's striving to reach the spirit of architecture.
>
> — Louis Kahn, 'Space and the Inspirations' (1967)

> 比起交談，我更喜歡繪畫。繪畫省時，且不會謊言連篇。
>
> ——勒·柯比意 (Le Corbusier)

對建築師而言，繪畫是最基本的交流方式。建築師、客戶、供應商之間的溝通就是以圖像作交流的。建築師們在醞釀概念初期，亦常會將不同的想法速寫於草稿紙上。於是，畫紙上便會重複出現各個想法的雛形、概念演變以及再生的痕跡，這是逾越感官的一種基本交流方式。當建築師完成初期的概念創作，進而開始更深入設計時，這些繪圖所發揮的功能與作用，則大不一樣。舉個例子，概念初稿僅注重整體架構，構圖方面相對而言較抽象，欠缺細節的描繪。初期的繪圖旨在為客戶呈現設計概念的基本雛形，讓客戶釐定設計方向，因此，概念形成時期的繪圖相對而言較簡潔。

> 建築師均以繪圖及設計致力達到建築的精神。
>
> ——路易斯·康 (Louis Kahn), 'Space and the Inspirations' (1967)

The Evolution of Drawing Practices

Today, the tradition of drawing on yellow sketch paper and tracing and retracing the master's work to develop a project has given way to contemporary digital design. Computer-aided tools have changed the culture and comprehension of architects' drawings in an unprecedented way. Three-dimensional modelling can be coupled with fabrication methods, so even tentative thoughts may appear to be rendered 'final'. These first drawings are meant to provide the client a means of cultivating the architect's ideas and spatial concepts. The craft and intelligence of executing new work is apparent in all drawing methods, whether generated by tedious hand tracing on vellum with technical ink pens and T-squares, or through the quick click of a computer mouse.

This chapter opens with discourse of the evolution and development of drawing practices. Through examples ranging from the old—construction drawings on vellum, personal sketchbook narratives, and abstract ink or charcoal paintings—to the new, including sophisticated parametric-formulated geometries, the array of works shown here not only reflect how architectural works are conceived and perceived but showcase their diversity and beauty which transcend mere function to become art.

Though it is inevitable that the old modes of drawing have been more or less replaced, there remain a few practitioners who are reluctant to surrender their T-squares and triangles. Some find themselves nostalgic for the meditative intellectual spaces that open up as they trace lines on paper. Like a sculptor working on a piece of wood or a musician stringing an instrument, an architect can yield to the joy of developing thoughts through the act of drawing and find emotional peace in its slow emergence.

An architect's repetitive tracing and retracing of lines amounts to more than a mere transfer of information but is in and of itself a meditative indulgence into the plan, a kinaesthetic reviewing of information. The pen in the hands of an architect is not qualitatively different from wood in the hands of a sculptor or an instrument in the hand of a musician. The drawing takes repetition to perfection; hence, architects love to trace and retrace their sketches to perfect the scheme and channel the vision from mind to hand.

In his recent book *Why Architects Still Draw* (2014), Paolo Belardi, an architect and professor in the Department of Civil and Environmental Engineering at the University of Perugia, wrote 'drawing by hand is an active way of thinking'. In our era where drawing software and AutoCAD are widely available and 3D-scanners and GPS devices are so close at hand, the question is: why do architects still draw by hand? Here, Belardi offers an elegant and ardent case for the value of the organic relationship that drawing creates between the mind and the hand without shunning new digital media for design.

Belardi suggested that drawing is the holistic manifestation of a particular design. 'It is the paradox of the acorn: a project emerges from a drawing—even from a sketch, rough and inchoate—just as an oak tree emerges from an acorn. Citing examples not just from architecture but also from literature, chemistry, music, archaeology, and art' (Belardi 2014).

Belardi further argues that 'drawing is not a passive recording but a moment of invention pregnant with creative possibilities'. Over the centuries, architects' drawings have evolved in style and philosophy, from Piranesi's eclectic etchings (1720–1778), to the modernist approach of training architects to develop ideas through the very act of drawing. This approach, while 'modern', was actually developed as early as 1937 by Walter Gropius, who initiated the then new notion of using drawing as a focus to train architects in the pedagogical program for his modern paradigm of architectural education.

This chapter consciously chooses not to dive into the discourse that drawing is the underpinning principle of architectural education but rather the simplest way to digest discussion on the approach of setting out to draw from a simple sketch on vellum to a full set of production drawings.

Those who appreciate Kahn's preference for sketching on vellum and his devotion to repetition watch raptly as his squares shift, turn, and interact in countless forms of spatially orientated variation. In the chapter 'Drawn on Yellow Paper: Toward a Culture of Lingering' (Merrill 2010), Kahn explains his preference for the thinnest of yellow paper: 'this is like a painter's flag, for work in progress, that however detailed and elaborate that which is drawn upon it may be in fact, still "wet", in flux, subject to change, criticism, rejection.'

This chapter will not be a full survey, nor will it indulge in the history and culture of architectural design and drawings; rather, it offers an overview of the development of a set of 'informed drawings' as well as things to consider when a beginner assumes the task of drawing. The discourse opens with the introduction of the various types of drawing and their trajectory, from the preparation of tender drawings, to detail drawings, and finally a set of construction drawings. The complete set of tender and construction drawings may convey similar information; however, depending on whether the drawings are intended for communication to potential tenderers or to the worker on site, they will cater to each accordingly. We would like to thank the architects who have contributed their drawings and images to this chapter as illustrations.

繪圖的進化史

將概念繪於黃色素描紙上不斷進行修改、潤飾的傳統設計方式，已不復存在，取而代之的是當今新興的各類電腦設計軟件。這些新興的電腦設計工具以前所未有的方式，改變了建築師們對於繪圖的理解及實踐。三維立體模型配以加工法，能夠將臨時性的設計想法表達得相當精細，所呈現的設計圖已似最後成品。然而，即使新舊設計方式所採用的平台及工具有所不同，兩者都能培養建築師的概念及空間感。不論是用專業墨水筆和丁字尺，不厭其煩地在牛皮紙上慢慢勾勒，還是簡便快捷地輕按電腦滑鼠，都能夠將創作一份新作品的工藝及智慧，詮釋得淋漓盡致。

本章會介紹大量的案例，以講解繪圖的進化史，從舊式畫在牛皮紙、素描簿上，以及用水墨或炭筆作抽象畫，到新式基於標準參數以繪製幾何圖形。案例中所涉及到的一系列作品，不僅為大家講解了建築的構思及感知，且詮釋了建築作品的多樣性及美感，令建築昇華為藝術品。

儘管舊的繪圖模式無可避免會被取代，但仍然有少數從業者不願意擯棄丁字尺和三角板。在紙上繪圖能夠勾起懷舊的思緒，沉醉於設計的空間中，好像一位雕塑家努力雕琢一塊木頭，或是一位音樂家細心彈奏一件樂器。通過繪圖這個動作，一名建築師總能夠在構思概念的過程中找到樂趣，並在概念漸現的過程中，找到內心的滿足感。保羅·比拉迪（Paolo Belardi）是佩魯賈大學土木與環境工程系的教授，同時作為一名建築師的他於新書《為什麼建築師們仍然作畫》（2014）中這樣寫道：「手工繪圖是一種積極的思考方式。在當今繪圖軟件廣泛應用、三維掃描儀以及全球衛星定位裝置唾手可得的年代，我們該問的是為何建築師們仍然會選擇手工繪圖？」比拉迪認為繪圖是一種直接的方式，讓大腦向雙手傳達訊息，而不用迂迴地透過電子媒介表達設計意念。

比拉迪認為圖則可以表達設計的整體性。「這種設計的過程就像橡果悖論一樣有違常理：一個項目始於一張圖，甚至有可能是一幅粗糙的原始畫作，就好比一棵橡樹始於一顆小小的橡果。」（Belardi, 2014）所引述的例子不僅僅涵蓋了建築學，且包含文學、化學、音樂、考古學以及美術。

比拉迪進一步論證繪圖不是被動地將想法記錄下來，而是一種「創意的厚積薄發」。在過去的幾個世紀裏，我們見證著建築師們創造出一系列非凡的作品，從最早期皮拉內西（Piranesi）不拘一格的蝕刻版畫，到以現代主義形式訓練建築師的繪圖技術及思考方式。沃爾特·格羅佩

斯（Walter Gropius）亦於 1937 年提出新的訓練建築師的方法，將繪圖作為現代建築學教育中的重點內容。

本章不會將重心放在「掌握繪圖技藝為建築教育的基本原則」這一話題上，而將著重探討由牛皮紙上的簡單素描，到最終完成的設計圖所採用的方法。

路易斯・康堅持在牛皮紙上重複地素描，在每個繪畫的角落變化著、互動著，以無數的組合嘗試不同的建築空間。在〈繪於黃紙上：走向勾留的文化〉一章中，路易斯・康這樣解釋他對最薄的黃紙的偏好：「未完工的作品，就好比是畫家的一面旗，不管該作品事實上已有多麼精細靈巧，它仍未『晾乾』，仍可改動，並隨時受制於外來的需求、批評甚至乎拋棄。」

一名建築師於畫紙上所勾勒出的原始線條，以及後期多次加工的痕跡，不僅僅傳遞著一種簡單的訊息，其創作本身便是一種認知的過程，一種對設計的深度反思。建築師手中的筆，從本質上相當於雕塑家手中的木頭，抑或是音樂家手中的樂器。熟能生巧，於是建築師鍾情於不斷修改和加工手中的繪圖，直至其完美。

本章的重點既非對建築繪圖史進行一場深入探索，亦非追溯其歷史與文化，而是提供針對一系列知情繪圖的概述，以及相關的初學者需知。

本章開篇會介紹不同類型的繪圖以及它們的製作方式，包括最初的招標圖則，到細節圖及最後的施工圖。一套完整的初期招標圖，則與最後的施工圖有可能非常相似，取決於圖則是否用於招標或施工參考。作者必須感謝為這章提供圖則和照片作插圖的建築師。

Tender Drawings and Contract Documents

Built environments are often complex, involving many designers, planners, consultants, and contractors in different fields and multiple stages of operation. The documents defining the contract are also complex and need to be comprehensive to accommodate the diversified backgrounds of each specialist. The task of preparing them for tendering therefore requires close attention to detail and a uniformity of approach, so as to achieve a coherent set of documents, forming an unambiguous and manageable contract. Aside from a series of drawings, a typical set of documents needed for tendering include the following:

- general conditions
- site preliminaries
- instructions to tenderers
- general and particular conditions of contract
- specification
- bill of quantities
- tender and appendices

For the built industry, the tender is a formal invitation to suppliers to make an offer to the buyer for the supply of goods and services as set out in the specification document within the formal tender document. In architecture, a contract is the formal agreement between the client and the contractor; it states the various services and deliverables expected of the contractor within the requested timeline and the expected contract price.

Contracts form an important part of the materials needed in calling tenderers to submit for a construction project. In the built/construction industry, the contract is a legally binding agreement between two parties with respect to the obligations of each party to the other and the liabilities.

There are several components to the formation of a complete contract.

招標圖及合同文件

建築環境是一個相當複雜的概念，因其涉及多個階段、多個行業的不同設計師、規劃師、顧問及承包商。正因所涉及的行業眾多，合同文件的條文則變得相當複雜，所涵蓋的內容亦因此必須相當全面。為準備該合同，創建者與相關協助者需十分注重細節以及方案的統一性，方能成功完成一份合理、清晰、可行的合同。除了一系列的繪圖，一套典型的招標文件還需包括下列內容：

- 一般條款
- 地盤的基本資料
- 投標指引
- 一般以及特別合同條款
- 規格/詳細說明
- 工料清單
- 標書與附錄

對於建築業界來說，投標是一份正式邀請函，邀請各供應商為某商品或者服務競價，具體競價商品或服務，則記錄於正式投標文件中的規格/詳細說明中。對建築界來說，合同指的是客戶與承包商在指定時間內提供指定價格的服務。

合同在一個工程項目招標的整個過程中，扮演著一個非常重要的角色。建築業的合同為雙方針對責任與義務所制定的合法並具有約束力的協議。

一份完整的合同需包括以下部份。

Instructions to tenderers

Instructions to tenderers inform the contractor where and when to deliver the tender as well as information regarding guarantees, bonds, and issuance. They may also contain information regarding items that will be supplied by the employer and sources of materials to be used in the contract, as well as proposed methods for construction.

投標指引

投標指引詳細說明了承包商該於何時何地投標以及所需相關保證書、債券、保險等信息。指引中亦有可能同時包括僱主所提供的特定條件、所採用的具體材料以及預計的施工方案。

General and particular conditions of contract

The general conditions of the contract may include any of the forms of contract. These may be amendments or additions that the employer wishes to make to the standard conditions. Standard conditions are not usually reproduced in the tender documents but will be named by specific reference, and a schedule will show the changes that have been made to them.

一般以及特別合同條款

一般條款可能包含任何一種「標準」形式的合同。而特別合同條款則由基於一般條款之上的修訂或附加條文組成。通常標準條款將不會在投標文件中再次被提及，只會以特定參考命名，且附有一份進度表列明上述條款的改動痕跡。

Specification

The specification focuses on quantitative information. It explicitly states the quality of materials and workmanship utilized, methods of installation, and finally, details regarding any laboratory testing to be used to comply with regulatory approval. The specification usually starts with a description of the works to be constructed, followed by all relevant data concerning the site, places of origin, size, the bonding method, and finishes. (Details are in Chapter 5.)

規格／詳細說明

規格／詳細說明中所涉及的主要是定量的資料。當中列明了材料的質量、所用工藝、安裝方法以及符合法例要求的各項詳細資料。規格／詳細說明通常先以工程介紹作開始，然後再逐一指定各種詳細要求，包括：工地基本資料、起源地、尺寸、結合方法、成品（其他詳細內容在第五章詳述）。

Bill of quantities or schedule of rates

The bill of quantities or schedule of prices is often interpreted as similar, but in fact the two are quite distinct. The Bill of Quantities (known as BQ) is frequently prepared by quantity surveyors. It is prepared for the costing of a building, and data are estimated from measurements provided by architects, structural engineers, and other building consultants. The costing information in a BQ contains estimates on various surface areas of the building in meters, including walls, floor areas, and roofs. It also contains a formal count on the number of doors, windows, and building service systems such as heating, cooling, plumbing, and electronics. It should be noted that the costs of labour and materials fluctuate constantly in the contracted period.

Historically, the practice of estimating building costs in BQ format arose from non-contractual measurements; the tenderers used drawings to assist in quoting lump-sum prices. A BQ shows the number or quantity of each item and its unit of measure, the rate per unit of quantity as quoted by the tenderer, and the consequent total price for that item. Bills may be quite complicated and contain hundreds of items, classified by trade or by a standard method of measurement. Other bills contain far fewer items.

Whereas the BQ provides an itemized list that includes the works to be constructed against each item of which the tenderer must quote a price, a schedule of prices may, by contrast, be much less specific. It may list provisional quantities, which are estimated. The schedule of prices is useful in many cases, including times when quantities (number of items, dimensions, or total areas) are fluctuating or uncertain.

工料清單及估價表

通常在合同中，一份工料清單可理解為一份估價表，但實際上兩者大不一樣。工料清單通常由工料測量師所制定。工料清單估算的是一棟建築的價格，這價格是基於建築師、結構工程師以及其他顧問的繪圖而估算出來的，其中包括牆壁面積、地面面積、屋頂面積、門窗的數目、冷暖氣供應設備、管道系統以及電器設備。不過，制定工料清單時要留意勞工成本與材料價格在合同期間的波動。

以往基於工料清單對建築成本的估算均為非合同類措施，投標者以建築繪圖作參考制定估算價格。一份工料清單中列明了每一個項目的數量及計算單位、每單位競標價格以及該項目的總價。工料清單有可能極其複雜，其中包括上百件項目，按交易分類或依照標準測量分類。而其他類型的清單相對而言較簡單。

估價表相對而言比工料清單簡單。承包商必須在工料清單中詳細列明每一個項目的工序及相應的工料及價格，但估價表只需列出估算的工料數量。因此在工程需求（項目所需的數量、尺寸及總面積）會變動或不明確時，估價表便會大派用場。

Tender and appendices

The tender is the tenderer's formal offer to undertake the contract; it is where the tenderer enters his or her sum price. The appendices to the tender contain other matters defining the contract's terms from which the tenderer will confirm acceptance of the offer. These terms may include a specific time for completion of the work, details of the consequences for failure to complete on time, the minimum amount of insurance, and the completion of bonds. Other specifics may include sources of materials and currency exchange rates (for international contracts).

標書與附錄

投標即投標人正式接受合同，在標書中清楚列明投標人所投總額。而交予投標人的附錄，則包含了合同條款的定義，投標人同時亦會參考該條款以確定是否接受招標邀請。條款可包括工程完工的具體時間、未於指定時間內完工的後果、保險金額的最小數目等等。此外，標書亦有可能列明其他相關的條文，例如貨幣兌換率（適用於國際性合同）或材料的來源。

Contract drawings

This chapter focuses on the complete set of contract drawings, which contain the entire scope of works involved. The more complete the contract drawings are, the more accurately the contractor can price the work, and the more easily he or she can avoid unexpected expenses during the later stages of the project. However, an extremely detailed drawing is not necessary at the tender stage (e.g., concrete reinforcement drawings). The contract drawings provided to the contractor should clearly show the requirements.

Depending on the complexity of the project, most jobs require at least two and sometimes three or more volumes of contract drawings to constitute a full set of documents. For small domestic renovation jobs or a design-and-build project, often one volume is sufficient for the contractor.

The overall objective for the main contract of a built project is to express the holistic voice of the architect. Furthermore, any documents included in this type of contract record every aspect related to the built form of the project, including site plans that outline everything from site boundaries, floor plans, sections, and elevations, all the way to the finer details of the stairs, roof, corners, and other interfacings of materials.

This chapter focuses only on the tendering of the main building. Descriptions of site-related work, such as site formation, demolition, and foundation are not covered.

合同圖則

本章重點在於介紹一套完整、涵蓋所有工作範疇的合同圖則。合同圖則越完整，承包商的估價便越準確，同時承包商亦能避免在工程期間的額外支出。然而，在投標階段的合同圖則並不需要達到最後合同般詳細具體，只要令投標商清晰明白合同要求即可以接受。

雖然每個項目的複雜度各異，但大多數項目均要求一套完整的文件中含有至少兩冊、有時達到三冊或更多的合同圖則。然而，對於本地的小型改善工程或設計建造項目，可能一幅圖則已經能清楚表達要求。

主合同的首要目標為詮釋建築師的主要設計意念，因此，主合同的文件必須詳細紀錄建築形態的各個層面及細節，其中包括了地盤界線、建築平面圖、剖面/截面圖、立面圖，及更具體的樓梯、屋頂、轉角以及其他銜接面等大量具體資料。

本章僅針對主大樓的招標，與工地相關的介紹內容，例如工地基本資料、拆除及建設等技術，將不會提及。

Taking the example of the Polytechnic University podium project designed by BARRIE HO Architecture Interiors Ltd., the plan (Figure 1) illustrates two levels of understanding: the illustration depicts the overall layout of the project, with the established grid set up at 9,000 mm apart in both directions with a clear location for the structure and expanse of the canopy. Through this detailed layout, one begins to understand not only the specific components of the structure but its placement and orientation in relationship to the surrounding buildings (Figure 2).

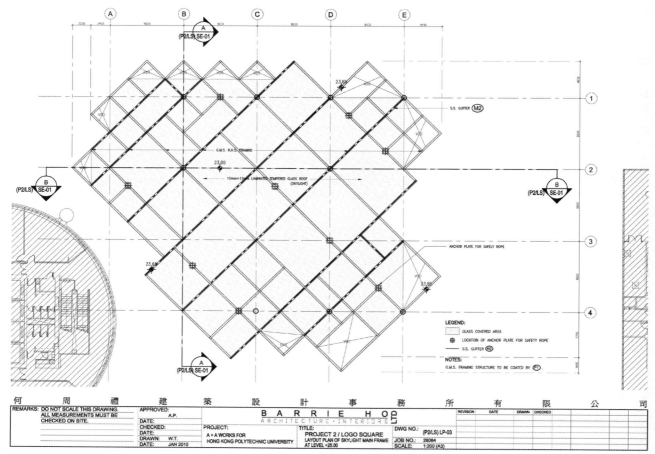

Figure 1: Logo Square layout plan of the Hong Kong Polytechnic University podium project

Figure 2: Glass canopy within the courtyard of the Hong Kong Polytechnic University

以何周禮建築設計事務所設計的理工大學平台項目為例，圖 1 表達了項目的兩個層次。第一，該圖則展示了項目的整體佈局，以 9,000 毫米的間隔在兩個方向上建立網格，清楚地標示天篷的結構和空間。通過該詳細明確的佈局，人們開始理解該結構特定的組件，甚至了解到其位置、座向及與周圍建築物的關係（圖 2）。

Building plans document the entire floor area. These plans should include details such as structural columns, lift cores, staircases, fire exits, doors, and windows. While the layout plan illustrates the essential content of the building, a site plan shows its surroundings, as seen in Figure 3, which shows the site plan from the Hong Kong Polytechnic University podium project. Typical components of the site plan may include adjacent roads, loading/unloading areas, existing planters, staircases, and neighbouring buildings.

建築圖則記錄了整個樓面的詳細資料，如結構柱、電梯、樓梯、消防通道、門窗等。佈局圖著重於建築的基本內容，而地盤平面圖則著重其附近的範圍。如圖 3 的香港理工大學平台項目所示，地盤平面圖的內容包括鄰近道路、裝載／卸載區、現有的植物、樓梯和鄰近建築。

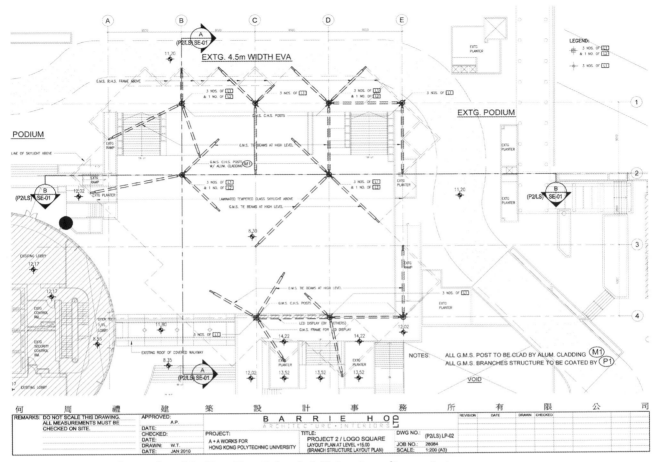

Figure 3: Logo Square layout plan of the Hong Kong Polytechnic University podium project

The clarity of the drawings stems from the consistency of the grid system, the relevance in scale, and the relationship of the proposed built object with the neighbouring buildings, so that when the drawings are referenced with the layout plan, one can see the 'additional information' drawn. The Polytechnic University podium project is rather abstract, as the final project is not a traditional building but a glass canopy that is housed within a university courtyard (Figure 4). Therefore, a section is essential in reading the full set of documents.

Sections are cut to reveal lift shaft construction, construction beams and slabs, and stairwell details. As it is difficult to document all the details in full, most of the tender drawing set only covers essential details, such as lift shaft construction and details regarding waterproofing and the roof. In Figure 6, the section of the built structure reveals that the form is supported by slender columns, while the canopy is supported by galvanized mild steel (GMS). Here, the grid system corresponds to the plan grid, so when the builder refers to Grid 2D in the plan, he or she can easily trace to the first column in the section drawing in Figure 6. Elevations are as important as any other drawing is, and document essential façades, such as window walls and interior partitions.

Figure 4: Glass canopy in the Logo Square of the Hong Kong Polytechnic University

Figure 5: Glass canopy in the courtyard of the Hong Kong Polytechnic University

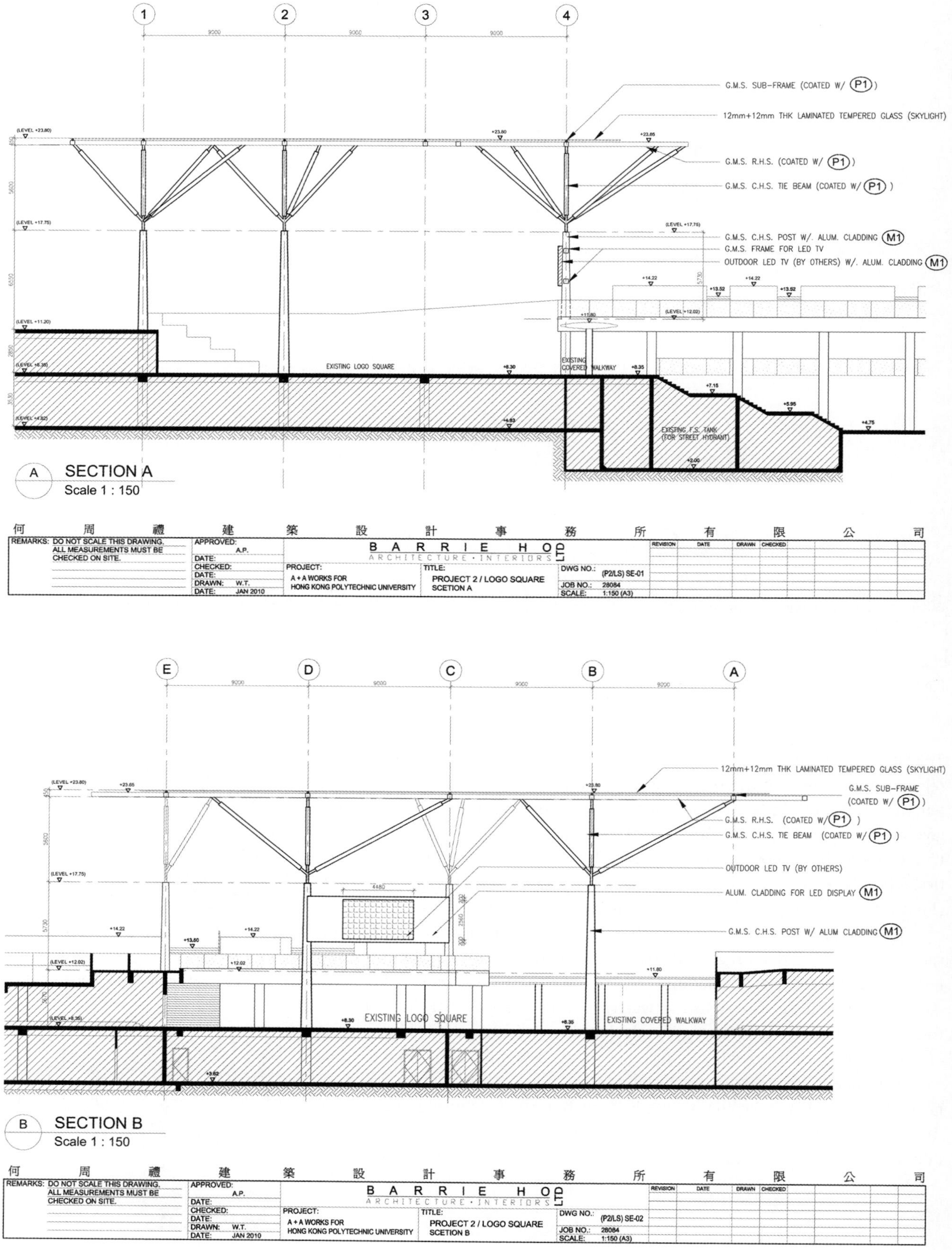

Figure 6: Section of the Logo Square at the Hong Kong Polytechnic University

清楚的圖則源自統一的網格系統。網格系統提供清晰的比例及設計項目與鄰近建築的關係。當互相比對其他圖則時，便可以清楚地看到「新增的資料」。在精心定義的網格系統下進行設計，可以方便修改，從而幫助在各種候選方案中，選擇出具有最合適佈局的方案。理工大學的這個項目，是一個比較抽象的玻璃天幕，並非一個傳統建築，因此需要附上橫切面圖，以令讀者明白此設計。

　　橫切面圖會顯示電梯井、樑柱、樓板和樓梯間的結構及細節。由於難以記錄所有的細節，大多數招標圖則只包括必要的細節，比如電梯井結構及有關屋頂的防水細節。在圖 4 中，橫切面顯示了玻璃天幕的形狀是以細長柱支撐，並以鍍鋅低碳鋼（GMS）支撐天幕。這圖則裏的網格系統對應平面圖的網格，所以讀者可以容易地比對在兩幅不同的圖則中的同一個結構，如第一條柱的部份。立視圖可以展示牆壁及室內分區，與橫切面圖同樣重要。

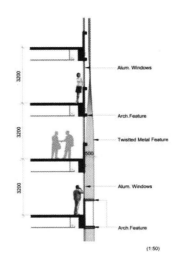

Figure 7: Section drawing of the twisted fin of GENESIS, Wong Chuk Hang Road

To illustrate the intricacies of an elevation, we use GENESIS, another project designed by BARRIE HO Architecture Interiors Ltd. which is located on Wong Chuk Hang Road. Figure 7 and Figure 8 illustrate how a side section and an axonometric drawing respectively of the building consist of critical information, such as the building's overall height, floor-to-floor height, window sizes, and the locations of any architectural features such as fins, eaves, and projecting window frames. Since the size of the elevation drawing is limited, there are several reference bubbles to lead the readers to an intricate set of drawings at a larger scale to more accurately reference the connection detail, fixing details, and material, as shown in Figure 9.

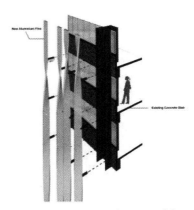

Figure 8: Axomoteric drawing of the twisted fin of GENESIS, Wong Chuk Hang Road

為了說明立視圖的複雜性，我們採用何周禮建築設計事務所設計位於黃竹坑道的 GENESIS 作另外一個例子。正如圖 7 所示，只是大樓的一側便包括了多個重要的資料，如建築物的總高度、樓層之間的高度、窗戶大小，以及任何建築特色，如垂直扭曲片、簷的位置和伸出的窗框（圖 8）。由於立視圖的大小是有限制的，因此需要多個參考的註釋，讓讀者去參考其他更大比例的圖則（圖 9）。

The most significant implication of the tender drawings is to attract potential tenderers to bid for the project based on the content, financial implications being a major aspect of the tender stage. Therefore, drawings should be as detailed as possible and reflect the final product as accurately as possible, to reduce the likelihood of misinterpretation and confusion, particularly in pricing. The only difference between a set of tender drawings and a set of construction drawings is that, once the tender is awarded, the drawings will become legally binding as part of the contractual agreement between the client and contractor. In a typical project built in Hong Kong, this official set of drawings is known as the AI-001; upon awarding the tender, the drawings will go to the contractor and become the instructions for development.

　　Construction drawing is a means of showing in graph form the technical aspects of the built work, including building size, relevant position of the building itself on a specific site, the composition of the materials to be used, and the method of construction.

　　Construction drawings not only include those produced by architects but also encompass structural drawings produced by structural engineers,

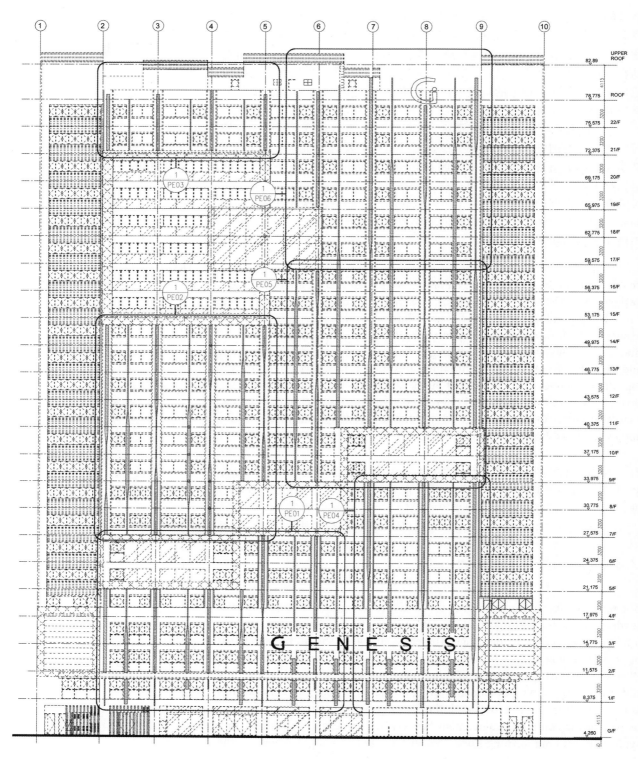

Figure 9: Front elevation drawing of GENESIS, Wong Chuk Hang Road

and building engineering services drawings, prepared by building services engineers.

招標圖則最重要的作用，是讓潛在投標商為項目作出估價。因此，圖則需要盡量詳細並充分反映最終成品，以便降低產生誤解或困惑的機會，尤其涉及到價格時需特別謹慎。招標圖則與施工圖的唯一區別，在於一旦招標已完成，相關圖則即刻成為客戶與承包商間具有法律效力協議的一部份。在香港，一個典型的工程項目所涉及的正式圖則，亦叫做 AI-001，其使用權如今歸承包商所有，同時亦作為建築指引。

通過突出尺寸大小、地盤位置、材料及建造方式等相關料，將建築工作中的技術性信息在施工圖上表達出來。

「施工圖」這個術語不僅涵蓋了建築師所繪的圖則，同時亦指結構工程師所作的工程結構圖以及屋宇設備工程師手中的屋宇設備圖。

Figure 10: Image of GENESIS, Wong Chuk Hang Road

Setting up a Drawing

To begin the drawings, the architect sets up and assigns the drawing grid, north arrow, title blocks, drawings title, and job number. These are the necessary steps to ensure the accuracy of the final product and smooth housekeeping procedures. The aim of all architectural drawings is to communicate the design concept and intent, as well as provide sufficient information for the contractors and suppliers to continue to carry out the scope of work required.

In addition to appropriately detailed drafting in CAD, attention must be paid to the fine details of the drawing: line weights, scale, dimension, and relevant notes. Each drawing should provide sufficient instructions to the builder at the given scale.

繪製一幅圖則

繪圖的第一步，是要準備繪圖網格、指北針、標題欄、圖則標題及工作編號。為保證最後成品的精準性以及便利將來的管理，上述均為必要步驟。製作圖則的主要目的，是明瞭設計意念，讓投標者及供應商獲得足夠的資訊，釐定相關的工作範圍及擬訂競標價格。

除了要注意 CAD 中精細的草稿圖，更要留心線條的粗細、比例、尺寸及相關註解。每幅圖則都應該在指定比例上為建築工人提供指示。

To illustrate the range of detail drawings in a set of tender drawings, we chose a project designed by BARRIE HO Architecture Interiors Ltd. The Community Service Centre for the Hong Kong Federation of Women (Figures 11 and 12) was constructed on a tight site under an existing flyover. Figure 13 shows the meticulous details embraced by this concept. The curved glass wall is composed of channel frames and faceted tempered glass with unconventional exposed fixings. As shown in Figure 12, the entrance detail also considered the limited site space and utilized tactful detailing such as the concealed external up-lighting, recessed door entry, and concealed I-beam structure, highlighting the entrance façade.

318 Drawing Practices: From Design Sketches to Tender Drawings

Figure 11: Community Service Centre of the Hong Kong Federation of Women under an existing flyover

Figure 12: Entrance façade of Community Service Centre of the Hong Kong Federation of Women

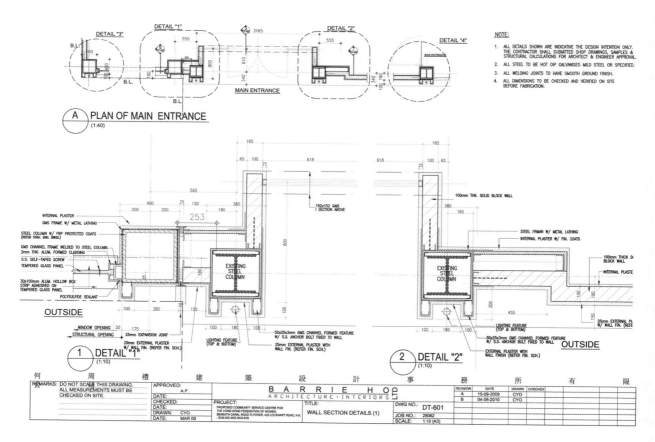

Figure 13: Detailed wall sections

我們以何周禮建築設計事務所設計的一個社區服務中心項目為例，說明招標圖則裏不同的詳細圖則。香港各界婦女聯合協進會的會址位於天橋下，受到很大的環境限制（圖 11 及 12）。圖 13 顯示這個會址的概念細節。建築師以鋼化玻璃及中通鋼架配以彎曲的玻璃來建立這非傳統的外牆。同時，入口的細節也考慮到有限的場地空間，利用隱蔽的向上照明、凹門入口和隱蔽的 I 型樑結構，突出入口的正面（圖 12）。

Each drawing sheet must also delineate the responsibilities of the involved parties: structural engineers, mechanical engineers, electrical engineers, quantity surveyors, and other consultants. Furthermore, the title block of each drawing sheet should indicate which party is responsible.

A complete set of working drawings usually presents building information in the following order:

- site plan
- floor plans
- framing plans
- roof plans
- reflected ceiling plans
- foundation/structure/framing plans
- exterior elevations
- building sections
- wall sections
- vertical transportation
- interior elevations/enlarged plans
- schedules
- details

每張圖則都須將工程所涉及對象的職責描述清晰，包括結構工程師、機械工程師、電力工程師、工料測量師等相關人員的職責。因此，每張圖則的標題欄需清晰指明責任承擔方。一套完整的工程圖一般遵照以下順序：

- 地盤平面圖
- 樓面平面圖
- 框架圖則
- 屋頂平面圖
- 反射天花計劃圖
- 地基圖則
- 室外立面圖／立視圖
- 截面圖／剖面圖
- 牆壁剖面圖
- 垂直運輸
- 室內立面圖
- 放大樓面平面圖
- 明細表或計劃表
- 細節說明

Site plan

The site plan presents the building site from a bird's-eye view and aims to communicate the context and surroundings of the project. It typically includes a heavy line denoting the building footprint but excludes a detailed depiction of the building itself. It is important to note that the architectural site plan differs from the landscape site plan. The former includes features of interest to the architect and engineer, such as lighting, electrical, fire hydrants, parking, ramps, paving, and property line, while the latter is used primarily by the landscape contractor.

地盤平面圖

為了清晰地表達項目的大致背景以及工地的周邊環境,建築師以鳥瞰的角度繪製出地盤平面圖。地盤平面圖通常只有一條勾勒出建築物大體輪廓走向的粗線,而不會細緻地畫出真正的建築物。值得特別一提的是,地盤平面圖有別於園景設計總圖,因後者主要由園林承包商所使用,而前者則需同時滿足建築師與工程師對於圖則的需求,包括照明、電力、消防栓、泊車、坡道、鋪砌面及界址線。

Floor plans

Floor plans are usually the first documents to be drafted in a construction project. Similar to site plans, they should be oriented so that the north arrow is pointed upwards or rightwards.

All plans should be drafted at the same scale (except for enlarged plans). Typically, the scale is 1:100, though depending on the size of the building, 1:50 or 1:200 may also be appropriate. The method of dimensioning carried out on floor plans should be consistent, as different methods may be applicable to a different project or building practice. The dimensions are referred to for construction purposes and for calculating construction costs.

Special attention should be paid to locating the structural elements as well as the building core, where vertical transportation (lifts, stairs) is usually placed. The architect will be equipped with knowledge of fire lifts, service lifts, or barrier-free lifts. In most circumstances, stair cores are grouped to allow maximum planning efficiency and will entail all the necessary support such as structural core, means of escape, fire hose reel, and pipe ducts. In domestic buildings, the kitchen and toilets must be located in the early phases to allow for provisions related to pipe ducts, and gas and plumbing requirements. For commercial buildings, the floor plans tend to have a central core with more open office plans to allow a more flexible layout, as office spaces tend to be partitioned into smaller rooms in later phases.

樓面平面圖

建築師通常都會優先完成樓面平面圖。樓面平面圖亦需要標明指北針。

所有的設計圖都該採用相同比例(除放大平面圖外)。通常而言,比例為1:100,具體情況仍需根據建築物的實際尺寸而定。此外,1:50或1:200的圖則亦同樣可以接受。必須統一不同的樓面平面圖中的尺寸標註,因為該圖則有可能適用於其他相關項目。尺寸會影響到建造方法及計算建造成本。

此外，我們還需考慮放置垂直運輸工具（電梯、樓梯）的建造槽。建築師掌握不同電梯的設備要求，包括消防電梯、維修電梯及無障礙電梯。一般而言，建築師都會將升降機槽放在一起，並將結構柱、逃生設備、防火喉及其他管道放在升降機槽附近，以提升空間的效率。在規劃住宅建築的空間時，首先要確定廚房與廁所的位置，以預留位置予煤氣與管道系統。而對於商業建築來說，樓面平面圖通常都會有一個中央核心的部份與開放式辦公室佈局，以允許後期更靈活地將空間分隔成更細小的房間。

Framing plans

A framing plan refers to the structural elements related to the concrete frames, including information on columns, beams, roof trusses, and slabs. The drawings are conveyed in sequence from structural elements to material finishes and communicated in a way that is immediately clear to the builders. While the drawings are likely to be prepared by structural engineers, the architect should ensure that the location of the walls is exact. There is, however, less control over the thickness of finishes, often resulting in less-than-precise room dimensions in these sets of framing drawings.

Last but not least, the cross-referencing of drawings begins with the floor plans. Since an effective set of working drawings means an extensive list of instructions for the builder, cross-referencing allows for elaboration on the pertinent drawing. To achieve this, drawings must be ordered hierarchically and include reference points from which builders can align various construction elements. For instance, column centrelines and designations play a central role in the cross-reference system. Since these are among the first elements to be built, they are used as a reference to locate other components of the building, such as the building core and interior walls. Other important cross-references include calling out section cuts, elevations, wall sections, and enlarged plans on the floor drawings, so that the corresponding sheet and drawing can be coordinated easily on site.

框架圖則

框架圖則通常用來標示水泥構架的資訊，其中包括樑柱、桁架及平板。框架圖則必須清晰簡潔，令建造者可以了解到由結構元件到材料飾面的不同要求。此圖則可以由結構工程師來製備，但建築師應確保牆壁的位置是精確的。不過，後續修建的飾面及批盪的厚度會影響到到房間的空間大小，令空間的精確度受到影響。

最後值得一提的是，圖則的交叉引用是從平面圖開始。一套好的工程圖會為建築工提供大量的資料，而完整清晰的交叉引用能夠利於理解相關圖則中的細節。因此，圖則必須以層次階級的方式排列，且納入參考註釋，以便於建築工人能夠與其他工程元素相對應。柱子的位置與中線在對應不同建築元素時是重要的參考，因為樑柱是最早被興建的一部份。其他可以用作位置參考的建築組件包括建築物的核心帶及室內牆壁。其他可以用作交叉引用的圖則包括截面圖、立面圖、牆剖面圖及放大平面圖。利用交叉引用法使相應的圖則能於實地得到較好的協調和參考。

Roof plans

A roof plan is usually drafted early on, not because it is needed in the initial stages of construction but because the information it provides is necessary for engineering consultants. Important elements on the roof include drains, skylights, lift machines, equipment housing, mechanical fans, and plumbing. Mechanical consultants need to locate HVAC units (fans, vents, cooling units) as well as evaluate the surrounding conditions in order to make necessary recommendations. Structural consultants have to take the slope of the roof into consideration when selecting types of waterproofing and accommodating drainage.

屋頂平面圖

建築師通常會預早準備好屋頂平面圖，這並非意味著工程初期就要用到它們，而是因為它所提供的資料，對於工程顧問而言十分重要。屋頂所涉及的重要部份包括雨水渠、天窗、屋宇設備、升降機設備、機械風扇及泵水系統。

技術顧問需找出適當空間放置暖通空調部件，且要視察並評估周遭環境的狀況，以便給予必要的改善建議。屋頂的排水功能亦十分重要，因此建築工程顧問需考慮屋頂的坡道設計及使用不同的防水物料。

Reflected ceiling plans

The RCP, a reflected ceiling plan, depicts the view of a ceiling if one were to look up from the floor of a room. Typical elements in the RCP are the location of lights, sprinkler heads, air-conditioning units, fire alarms, security cameras, smoke detectors, motion detectors, and other suspended elements. Major elements such as lights and HVAC will be affected by any alterations to the design of the floor plan, as they occupy larger areas of the ceiling. When designing the RCP, the architect must first locate structural members, then the mechanical and electrical requirements of the room. There are also minor elements such as access panels for repair and air inlet and outlet, which also need to be considered.

反射天花計劃圖

反射天花計劃圖其實就是樓面平面圖的鏡面反射。反射天花計劃圖將在房間向上望的不同元素都標示出來，包括燈、消防花灑、冷氣機組、火警鐘、閉路電視錄影鏡頭、煙霧感應器、活動感應器及其他懸掛的裝置。建築師會優先處理一些主要的構件如照明及空調，這些元素會佔據較多的空間，若樓面平面圖有任何的改動會造成更大的影響。

建築師在繪劃反射天花計劃圖時，先要安置大廈的結構性部件，再考慮機電要求及其他較次要的元素，如維修通道和空氣出入口。

Foundation plans

The foundation plan, similar to the framing plan, is a section of drawings that provide detailed instruction for the structural framework. These drawings are often drafted with the assistance of engineering personnel or consultants, and it is essential that architects work closely with engineers to identify all the features of the foundation. The foundation plan directly responds to the soil condition and is a set of drawings to be cross-referenced with site the formation plan. The type of foundation, whether it is mini-piles, H-piles, or large-diameter or small-diameter bored piles, will be determined by the engineer, depending first on the soil type. Along with depth of foundation, the engineer will determine how the foundation will sit on the bedrock, while considering sizes of footings and columns above the superstructure.

Figure 14: Conference Lodge of the Hong Kong University of Science and Technology

地基圖則

地基圖則與框架圖則相當相似，建造框架便是基於地基圖則，因此地基圖則需要相當詳細。一般而言，建築師會與工程師合作繪製地基圖則，以便凸顯地基的特點，如土地情況。地基圖則同時是一套重要的交叉比對圖則。工程師會就不同的土地情況、地基深度、如何打在石床上及地基的寸尺等因素，去決定使用不同種類的地基，如微型樁、工字樁、大小直徑鑽孔灌注樁等。

Elevations

Exterior elevations reveal a different layer of information from floor plans. Elevations typically show the profile of a building, showing the building materials, locations of windows, and building enclosure (Figure 14). Elevation drawings, like floor plans, are submitted to the Buildings Department for approval. The elevations can reveal information such as the size of the windows, locations of openable windows, façade material treatment, curtain wall sizes, and ways of fenestration—elements which cannot be shown on a floor plan.

室外立面圖/立視圖

相對於地基平面圖，室外立面圖/立視圖所呈現出的是另一層面的資訊。立面圖通常顯示建築物的輪廓、建築材料、門窗位置和建築物的外殼（圖 14）。立面圖像平面圖一樣，是提交屋宇署審批的圖則。立視圖能顯示平面圖不能表達的資料，如窗口的大小、外牆材料、幕牆大小和開窗方法。

In order to draw an elevation, one looks at the building from the front, taking into account the ground datum, the overall height of the building, the floor-to-floor height, and any features and openings on the façade. It is often more applicable to label any building materials on the elevations rather than on the floor plan.

Figure 15 shows an example of an elevation drawing in which the entire aluminium curtain walls where the various materials on the façades are is represented by a different hatch. This is indicated in the side legend.

建築師在繪劃立視圖時，先以一個正面的角度觀看建築物，考慮地面基準、建築物總高度、樓宇高度、立面上不同的特徵和開口。立面圖往往比平面圖更適合於標示出所用的建材。

圖 15 展示了一幅鋁外幕牆的立面圖，幕牆上所用的不同物料已在側注中指示出來。

Figure 15: Elevation of an aluminium curtain wall

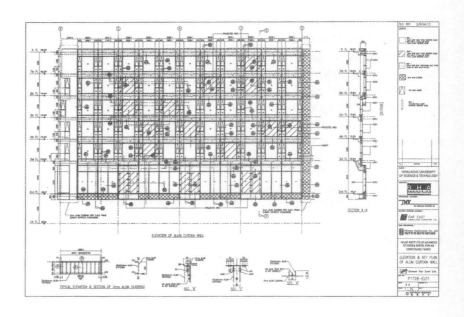

Building sections

A building section is a cross-cut in the middle of the building, revealing the internal elements, finished floor heights, staircases, and materials. Building sections aim to communicate floor-to-floor relationships as well as the construction of the building envelope. Most of the internal structures, including walls, floors, beams, slabs, joists, and blocking, will be shown in a building section drawing. Building sections are usually drawn at the same scale as are exterior elevations and floor plans, so that cross-referencing between drawings can be conducted easily. Through a properly drawn section, an architect can illustrate how a building sits on the ground, whether the building is a subterranean structure or is supported by columns or stilts. Therefore, building sections are an informative set of drawings to explain the relationship between the building and its immediate context.

截面圖/剖面圖

截面圖/剖面圖透出的建築內部元件，包括完成的地板高度、樓梯和建材。截面圖/剖面圖可以傳達樓層之間的關係以及建築結構的施工。大多數的內部結構，包括牆壁、地板、樑、板、托樑以及阻塞，都是截面圖的內容。截面圖、立面圖和樓面圖通常都是以同樣的比例繪製，以方便交叉引用。建築師透過截面圖/剖面圖去表達建築物如何建在地面之上，以及建築物的地下結構應由圓柱或高蹺支撐。因此，截面圖/剖面圖是一套最能表達建築物與周邊關係的圖則。

Wall sections

The drafting of wall sections requires a more thorough understanding of building assembly than do major construction drawings. The purpose of the wall section drawing is to provide the builder detailed instructions on wall assembly. Wall sections should be drafted at a legible scale, typically 1:10, and clearly convey the walls' structural elements, air space, and material finishes. Figure 16 depicts a project designed by BARRIE HO Architecture Interiors Ltd at the HKUST Jockey Club Institute of Advanced Studies and Hostel for

IAS Conference Lodge. This shows the components of a wall façade. The wall section depicts the threshold conditions showing both the interior and exterior of the wall and how the vertical elements such as spandrel, thermal insulation, and aluminium acoustic panels relate to the floor slabs. The Fire Resistant Construction Code requires a spandrel of 900 mm as an external wall separation between floors.

牆壁剖面圖

繪製牆壁剖面圖前，必須深入了解建築物的構成。牆壁剖面圖的目的，是指導建築工人建造牆面並註明特別要注意的資訊。牆壁剖面圖一般採用 1：10 的比例繪製，並清晰突出結構性元素、空隙及牆壁的飾面。圖 16 以何周禮建築設計事務所設計的香港科技大學賽馬會高等研究院及宿舍作為例子，展示了一些建築立面的結構。牆壁剖面圖顯示了一道牆壁的門檻狀況，以及不同結構與地板的關係，包括內牆及外牆的細節，其他垂直的結構如拱肩、隔熱層板及鋁製隔音板。耐火結構守則指明需要一道 900 毫米闊的拱肩作外牆來分隔不同樓層。

A typical exterior wall section drawing would also indicate details relevant to moisture and thermal protection, such as flashing, drip edges, seams, joints, insulation, and vapour barriers. More detailed drawings are required to illustrate the intricate details of building assembly for better clarity of the hierarchy of structure to be shown. Figure 17 is a typical detail of a 1:2 scale showing the construction components of a curtain wall. Each element is drawn in such detail that it is prescriptive to the way the contractor would construct this window. Later in the chapter, we will explore the specifics of the making of the construction details.

對於室外牆壁剖面圖，圖則應指出有關保濕及保溫的建築物細節，例如防水板、滴水檐、接縫、關節、絕緣措施及防潮層。很多時候都需要更詳細的圖則，說明建築組件的複雜細節。圖 17 展示一張幕牆一比二的的細節。這張圖則清楚指示承建商如何建造這個窗戶每一個不同的組件。

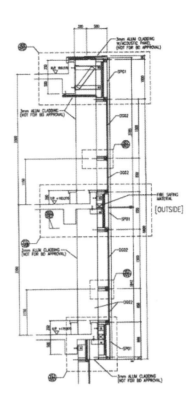

Figure 16: Vertical section of an aluminium section wall

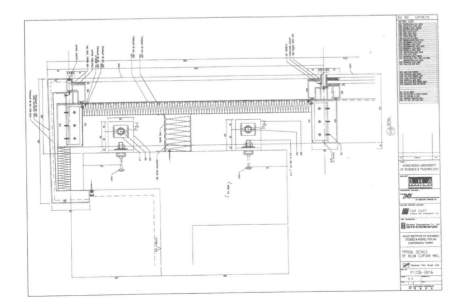

Figure 17: Typical details of an aluminium curtain wall

Vertical transportation

Vertical transportation broadly covers all lifts, stairs, and escalators, as well as ramps, ladders, and other elements that allow access between levels in a building. Since vertical transportation plays a central role in any building providing access and egress, and because building codes often require separate assessments of each of these elements, their respective plans, sections, and details have to be drafted sequentially.

The plans for vertical transportation are submitted to the necessary authorities for review and approval. It is important to note details for the guardrails and handrails, as well as the tread and raise dimensions. Escalators, though typically supplied by manufacturers, should be included to account for the detailing that will be necessary at the top and bottom of each run as it meets the floor slab and other finishes.

垂直運輸

廣義上，垂直運輸涵蓋了所有的升降機、電梯、扶手電梯、坡道、樓梯以及其他令使用者於建築物樓層之間移動的設施。一座大廈的垂直運輸是其核心部份，且建築條例一般要求對垂直運輸進行分開評估，因此，繪製此類設計圖、剖面圖及其他細節，應相當清晰有序。

為達到法例要求，相關的政府部門會評估及審核一座大廈的垂直運輸設計。因此，樓梯的護欄、扶手、樓梯踏板及尺寸必須清楚註明。儘管扶手電梯一般由製造商提供，但亦應該考慮其尺寸，以適當地安排樓層間的平板及批盪尺寸。

Interior elevations/enlarged plans

Enlarged plans and interior elevations are typically drawn at a 1:50 or a 1:25 scale, depending on the size and detail that need to be represented. The drawings focus on material finishes, case work, and interior detailing, the north elevation typically being the first in sequence and the rest proceeding in a clockwise order. Although not submitted as part of the Buildings Department's drawing set, the purpose of the interior elevations is intended for the interior designers in the making of the interior space.

室內立面/立視圖與放大樓面平面圖

室內立面/立視圖與放大樓面平面圖通常採用 1：50 或 1：25 的比例，具體取決於其所代表的尺寸與細節。此類設計圖的重點在於凸顯飾面、個案設計及內部細節設計。北立面通常為第一張圖，剩餘幾面則按照順時針方向逐步展示。室內立面/立視圖與放大樓面平面圖並不會交予屋宇署審批，這組圖則主要是讓室內設計師進行他們的工作。

Schedules

Schedules are compendiums that illustrate different types of window, door, and finish in the building. Each of these elements is given a designation, and their corresponding elevations and dimensions are drafted here. Included also are specifications on materials and hardware, so that quantity can be determined.

Elements such as window frame or door jamb requirements are further called out in the details section.

Window and door schedules are among the most tedious yet essential to develop, as each window of the building has to be encoded with a number to identify its property in glazing type, fixing detail, and size. This set of informative drawings gives the contractor a systematic method to reference the specific types of window chosen, allowing for a smooth execution during installation.

There is a long list of schedules such as door schedules, window schedules, ironmongery schemes, sanitary fitment schedules, and more, but the ultimate purpose of the schedules is to provide a matrix for contractors to easily see the specific requirements relating to the built item. Take the example of a door. The schedule will clearly indicate whether the door is a fire door, whether or not it has a vision panel, and whether the door swings left or right; finally it will indicate locations of hinges and handles. All of these elements should be easily referenced if the schedules are intact and properly executed. Figure 18 is an example of a door schedule with relevant numbers and specificities associated with the item.

明細表或計劃表

計劃表是一個綱要，說明建築項目內不同的門、窗類型以及牆壁的各種飾面。計劃表會指明每一個門窗和飾面的編號，附上相關的立面草圖和相應的尺寸，以及使用的材料和硬件規格。其他元件如窗框或門框的詳細資料，亦會在計劃表中列明，令承建商可以估算所需的物料。

門窗計劃表是最乏味但亦是最重要的清單。計劃表會編排窗戶的編號，清楚列明每一個的玻璃類型、安裝細節及尺寸。這種資訊性的圖則，令承建商能有系統地安裝不同的建築部件。

計劃表有林林總總的種類，包括門、窗、五金和廚廁設備等等。計劃表的最終目的是提供一個圖表，讓承包商可以很容易地看到有關該項目的具體要求。以門的計劃表為例，會表清楚地表明該門是否防火門、有否窗戶以及向左或向右擺動。最後，它會顯示門鉸和把手的位置。若果建造計劃能依據計劃表完整地執行，這些不同的部件都應和計劃表對照。圖 18 是一個門的計劃表，表上顯示不同門的相關數字及門的特點。

Construction details

There are several purposes for including such an extensive set of construction detail drawings in a given project. First, it allows for the accuracy of financial estimates, as going over budget is a legitimate concern. Second, well-designed drawings are integral to ensuring the quality of the final built project; well-executed drawings allow the architect to hold the builder to a higher standard.

Additional details are often called out and indicated on the other detail drawings. Each sheet may have relevant details to include, for example, the building envelope, walls, windows, doors, ceiling, and stairs. The scale of a detail drawing depends on context, but 1:5, 1:2, and even 1:1 drawings are sometimes necessary if the construction requires meticulous attention.

It is not uncommon for builders to demand additional fees when design details are added at a later date after estimates are made and contracts drawn. Although contractors do provide shop drawings for approval, details are often overlooked at this stage, and litigation involving design changes and payment will almost always delay the project and pose a liability risk.

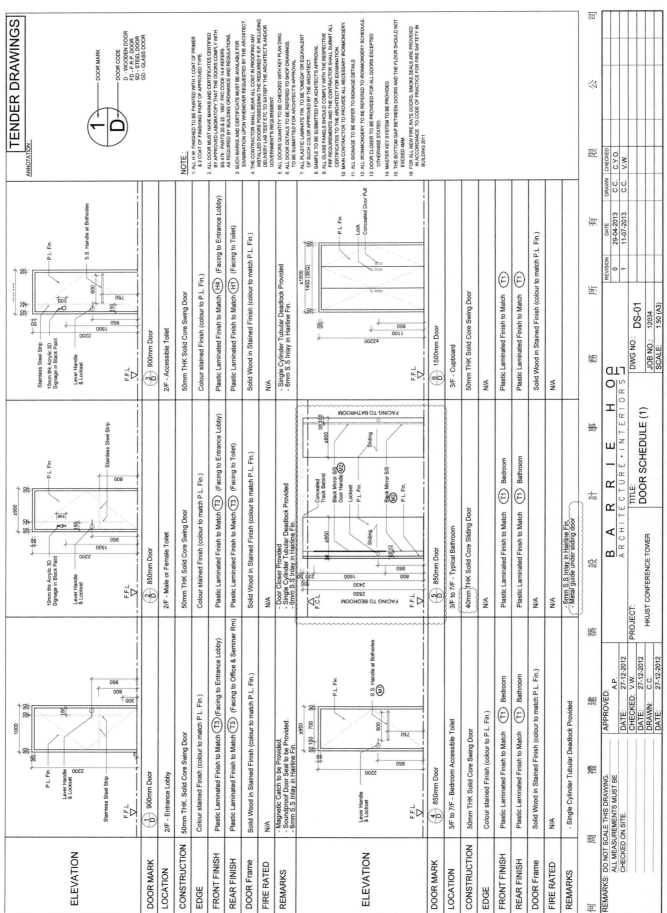

Figure 18: Door schedule

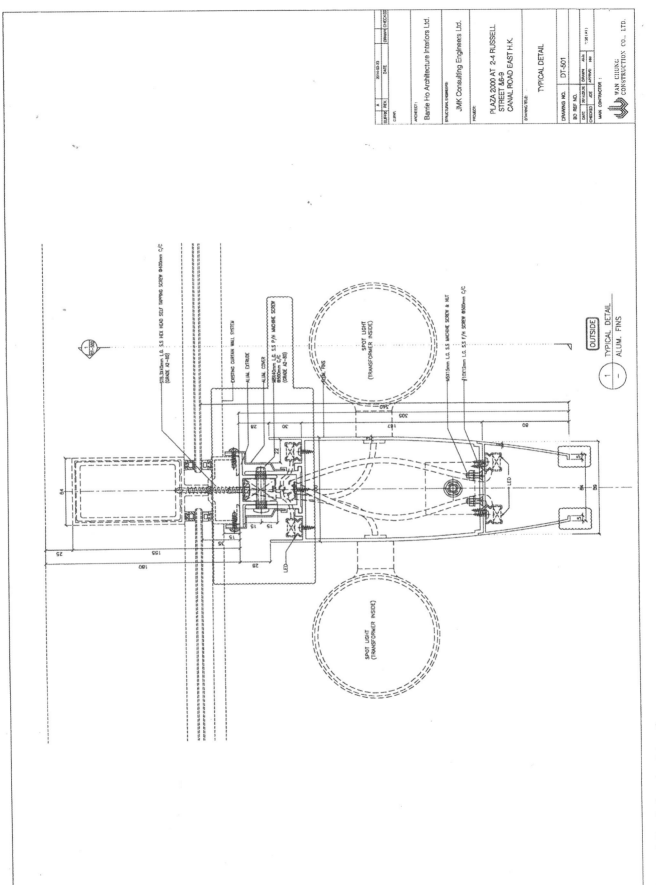

Figure 19: Detail of an aluminium fin

Figure 20: Plaza 2000, showing flashing, lighting, and vertical fin features

構造細節

為一個工程項目創作如此大量的圖則有幾個原因。其一，圖則有助於提高工程投標估價的準確度，將項目預算控制於合理範圍。其二，一套好的設計圖則能夠為建築工人提供高水平的參考，保證最終的質素。

工程圖則的補充信息通常會繪製於其他主要圖則中。每張圖紙都包含相關細節，例如建築物內外結構、牆壁、窗戶、門、天花及樓梯。若需格外注重工程的施工條件，1：5、1：2 甚至 1：1 的比例有時亦會被採用，但比例則取決於具體情況。

通常在估價結束且已經簽署合同的情況下，添加額外的設計細節將會增加額外費用。即使承包商的確已提供施工圖供批核，不同的施工及設計人員往往會在此階段忽視很多細節，且涉及設計圖和交易變動的訴訟，大多數情況下都會延誤工程進度，並有機會連帶引致負上法律責任的風險。

In essence, construction detail drawings offer a common graphical language whereby the architect and the contractor are able to communicate the minute details of a building's assembly. This common language is achieved by implementing a standardized presentation style with respect to lines, hatching, and symbols. For instance, in Figure 19, there is a consistency in the language of hatching to show components such as flashing, brackets, screws, and glazing, showing how a lighting feature and vertical fin are to be assembled at the building exterior. The drawing content of the fin detail shown in Figure 19 can be seen in the external façade of a building shown in Figure 20.

工程設計圖本質上就是為工程隊伍提供一套共同的圖像語言，令建築師及承包商能了解一幢建築物的細節。這套語言是通過訂立統一及標準的線條、剖面及符號，來確保不同的從業員能互相參考。以圖 19 為例，圖中的建築外牆部件如燈光、支架、螺絲和玻璃，都是用一個統一的圖像語言來標示。圖 20 的建築立面就是圖 19 所顯示的結果。

Conclusion

This chapter presents an overview and the hierarchy of a set of drawings. The way an architect organizes a set of drawings reflects his or her ability to compose a well-orchestrated picture of how a building is assembled. Therefore, other than clarity and consistency, an understanding of assembly and the role of technology play a determining factor in the way a full set of drawings is presented.

While this chapter offers a step-by-step trajectory through the execution of a set of construction drawings, it is not meant to be all-encompassing; rather, it attempts to convey the unique significance of each drawing, as well as the relationship between them, all with the ultimate goal of successfully realizing a concept from paper or computer screen to built form.

Many great architects, such as Carlo Scarpa in his Castelvecchio Museum in Verona, Italy, or Peter Zumthor's Val Thermal Bath in Switzerland, have utilized details as the driving forces behind a rigorous design concept. The architect is able to make a strong statement in his or her work, not only with the interplay of light and shadow, choice of materials, or how the building sits in its context and reacts to the building technology, but also how the construction details are manifested in the buildings. It is truly inspiring when an architect is able to push the building details beyond a mere construction process into a unique design statement.

This chapter presents only the beginning of a discourse on the various aspects involved in the practice of architectural drawing. It attempts to illustrate how a fully comprehensible set of construction drawings has the ability to transform a series of two-dimensional drawings into a three-dimensional built form. Through a detailed overview of the process of composing a set of construction drawings, this chapter provides a window into how architecture is conceived and consequently how it might be perceived.

Our contemporary practice of architecture today combines both traditional methods of drawing and those of high-tech digital design. Although some computer-aided tools have revolutionized the culture of the built industry and the comprehension of architects' drawings, the same fundamental communication is conveyed, whether it is between the architect's mind and hand during the initial conception of ideas, or between various engineers, consultants, and contractors in later stages of development.

As Architect Charles Eames once said, 'The details are not the details. They make the design.'

結語

本章提供了一個有關不同層次圖則的概述。一個建築師組織圖則的方式，反映了他繪畫一幢精心策劃的建築物的能力。因此，為了表達一套完整的圖則，除了清晰的圖則和一致性的圖像語言外，了解如何組裝不同組件及不同的科技，也有著決定性的因素。

雖然本章提供了製作不同施工圖則的過程，但並不是包羅萬象的，而是嘗試傳達每一種圖則的獨特意義，以及圖與圖之間的關係。圖則的最終目標都是令建築師能夠把紙上或電腦熒幕上的設計概念化為屹立的建築。

偉大的建築師如卡羅・斯卡帕（Carlo Scarpa）設計的意大利維羅納城堡博物館（Castelvecchio Museum），或彼得・卒姆托（Peter Zumthor）設計的瑞士瓦爾溫泉浴場，都能以嚴謹的細節驅動其設計理念。建築師能夠在作品中強烈地表達他們的理念，不止於玩弄光影、選擇建材、配合附近的環境、利用不同的興建技術，更能在施工的細節上表現建築師的理念。當一個建築師能夠利用不同的施工過程去設計建築細節，他的作品便逾越了單純的施工過程，成為一種別樹一幟的風格。

本章介紹了不同種類的建築圖則及其相關的應用。透過這個章節，讀者能夠了解到如何將二維的建築圖則轉變成立體的建築。通過對施工圖則的詳細介紹，本章還讓讀者了解到建築由構思到建造需經歷的過程，以及建築是如何「閱讀」的。

當代的建築行業結合傳統的繪畫及高科技繪圖，已經徹底改變了建造業及建築師對圖則的理解。無論是建築師在初步構思過程中的以手畫我心的表達，或各種工程師、顧問和承包商在發展後期的溝通，繪圖的基本理念是恆久不變的。

正如建築師查爾斯・埃姆斯（Charles Eames）所言：「細節並不止於細節，細節即是設計。」

References

Belardi, Paolo. 2014. *Why Architects Still Draw*. Cambridge, MA: The MIT Press.

Industrial Centre, the Hong Kong Polytechnic University. 'Reading Materials for IC Training Modules: Construction Drawing Practices'. Retrieved 6 March 2015, from http://www.ic.polyu.edu.hk/student_net/training_materials/IC%20Workshop%20Materials%2009%20-%20Construction%20Drawing%20Practices.pdf

Merrill, Michael. 2010. *Louis Khan: Drawing to Find Out. The Dominican Motherhouse and the Patient Search for Architecture*. Baden: Lars Müller Publishers.

Acknowledgements

I would like to thank BARRIE HO Architecture Interiors Ltd. for providing photos and drawings in this chapter.

Chapter 5
The Importance of Construction Specifications

施工規格的重要性

Wai Chui Chi, Rosman 衞翠芷

The Role of Specifications

Construction specifications are important documents that communicate the design intent of the architects and the engineers to the contractors. They are written descriptions of the project and are used to govern the quality performance of its construction works. The specifications, together with the agreement, conditions of contract, drawings, and the bills of quantities (BQs) in traditional design-tender-build projects, are major components of tender documents for bidding on the construction of projects. These documents will subsequently become contract documents for construction work after signing the agreement, which is the contractual agreement between client and contractor that the former will pay the contract sum when the latter performs the construction work as specified.

During the construction stage, specifications are essential documents for contract administration. The specifications spell out clearly the acceptance standards of the architect or the engineer on materials, workmanship, and other obligations under the contract. They are also important documents in determining any variations from the contract, measuring interim payments and imposing penalties for substandard work. In the final account stage, specifications provide a strong reference in ascertaining variation costs and claims, as well as settling disputes in contracts.

While drawings are graphic descriptions of the work to be performed by the contractor, specifications are the text description of it, and the BQs detail the quantities and cost. These documents complement each other and are essential for the complete understanding of the work. Whenever there are discrepancies among these documents, the contractor should seek clarification from the architect or engineer, who will clarify his or her intention through the instructions of the architect or engineer.

施工規格的作用

施工規格是建築師和工程師用來傳達設計意圖予承建商的重要文件。它們是項目的文字描述，用以管理建築工程的質量。在傳統的設計－招標－建造項目的招標文件中，施工規格，連同協議、合同條件、建築圖紙和工程量清單，是投標文件的主要組成部份。而這些文件在業主與工程承建商簽署協議後，將成為合同文件。簽署協議，代表著業主承諾當承建商完成執行規定的建設工程後將支付合同款項。

在施工階段，施工規格是合同規管的必要文件。該規格註明了建築師或工程師對材料、工藝和一般須遵守規範的明確驗收標準。它們也是用來決定何為合同中的改動，和計算中期付款，以及當工程質量低於合同標準時，計算罰款款額的重要文件。在決算階段，施工規格更為確定合同中的變動成本及索賠，以及在解決合同爭端中，提供重要的參考。

圖紙是用圖像，而施工規格卻利用文字，向承建商描述合同裏的工程，至於工程量清單則詳盡地列出工程內有關材料與工藝所需要的數量和成本。這些文件對理解整個工程內容是相輔相成、缺一不可的。每當這些文件之間的資料有差異時，承建商須向建築師或工程師查明，讓他們透過建築師指示或工程師指示澄清設計的本意。

Format of Specifications

Most common construction specifications used in Hong Kong are of the prescriptive type. Some may use the performance type of specifications, or a combination of both prescriptive-based and performance-based specifications. In the prescriptive specifications, all requirements of the work in materials, workmanship, and other obligations are specified in every detail. In the performance specifications, only the functional performance requirement of the completed work is specified, so the contractor is free to develop his or her own method to achieve the result. Prescriptive specifications give the architect or engineer more certainty of the final outcome of the construction and make process-control during the construction possible, but his or her knowledge of the construction activities is required when drafting the specifications. Performance specifications are more suitable for construction works that require the contractor's input in a portion of the design work, which is associated with construction technique, or works that call for the contractor's innovation in construction approach.

Different construction specifications are used for different contract types, including site investigation, site formation, foundation, superstructure, nominated subcontracts, etc. The specifications usually contain two parts: one is the general specifications, which are standard requirements and conditions of construction. They are usually used for all similar contracts in the same architectural or engineering firm. The other part is the particular specifications, which are specifications clauses particular to the contract works, and can take precedence over the general specifications.

施工規格的格式

在香港，最常使用的施工規格是規範類的。有些會使用工作性能類的規格，或集合規範類和工作性能類為基礎的規格。在規範類的規格中，材料、工藝和一般須遵守規範裏的所有要求，均詳細列明。然而，在工作性能類的規格裏，只須指明工程在完成時的工作性能要求，承建商可以自由地選擇自己的方法來完成工程。規範類的規格讓建築師或工程師更能掌握工程落成時的效果，並能控制施工過程，但它要求建築師或工程師對各項建築活動有所認識，才能起草施工規格。而工作性能類的規格，比較適合一些需要承建商投入與他們施工工藝相關的設計工作的項目，或一些要求承建商在施工技術方面有所創新的工程。

不同的合同類型會使用不同的施工規格，包括現場勘察、地盤平整、基礎、上層建築、指定分包合同等。施工規格通常包含兩個部份：第一部份是一般規格，這是一般標準的要求和建築條件。它們通常適用於同一建築或工程公司所有類似的合同。另一部份是特殊規格，這是特別為該工程合同訂定的規格條款。而特殊規格的要求是可以凌駕於一般規格的要求的。

Organization of Specifications

Most of the specifications are arranged in sections by construction trade in the chronology of the works, but some specifications may be arranged by the building components, for example, windows and doors, in which all trades of the same components grouped together. Specifications arranged by trade is more convenient for subcontractors of each trade to find their own work scope in the specifications. Arrangement by components has the merit of grouping all necessary trades in building the components, thus making it an easier reference for the manufacturers and subcontractors of the building components.

For specifications arranged by trade, both the general specifications and particular specifications are arranged according to the working sequence of building works on site. Take specifications for building works as an example. The typical arrangement of the various sections is as follows:

1. demolition
2. earthwork
3. concrete
4. masonry
5. structural steel
6. roofing
7. carpentry and joinery
8. ironmongery
9. curtain walls
10. metal works
11. finishes
12. sanitary appliances
13. glazing
14. painting
15. internal fittings and fixture
16. plumbing and drainage
17. landscape works

For complex projects, separate contracts are used for site formation, demolition, foundation works, interior decoration, and so on. Like the building contracts, the requirements for materials, workmanship, and other obligations of these contractors for the individual works are spelt out clearly in the specifications of these contracts. Similarly, the nominated subcontracts of the building service works, such as air-conditioning installation, fire services installation, water services installation, and electrical installation, will have their own specifications in each nominated subcontract.

施工規格的結構

大部份的施工規格，都是根據各建築行業在工程進度中出現的時序來編寫的。但有一些施工規格，則是使用建築構件來分類的，例如窗戶、門等，把所有製造該建築構件裏的各建築行業組合在一起來編寫。根據建築行業編寫的施工規格，有利於每一個行業的分包商，讓他們在施工規格裏更方便地找到自己的工作範圍；而使用建築構件來分類的施工規格，則方便那些建築構件製造商和分包商。

以建築行業為本編寫的施工規格，無論是一般施工規格或特殊施工規格，都會根據建築工程在現場工作的時序來編寫。以建築工程的施工規格為例，各部份的標準安排如下：

1. 拆卸
2. 土方
3. 混凝土
4. 石工

5. 結構鋼鐵
6. 屋頂
7. 木工和細木工
8. 小五金
9. 幕牆
10. 金屬
11. 塗飾
12. 潔具
13. 玻璃
14. 油漆
15. 內部配件及固定裝置
16. 給排水
17. 園景工程

對於複雜的工程項目，通常會把地盤平整、拆卸、基礎工程及室內裝修等工程，分開使用個別獨立的合同。但一如建築工程合同，這些合同也必須註明對材料、工藝和一般須遵守規範的要求。同樣，對於屋宇裝備的指定分包合同，如空調安裝、消防安裝、供水服務安裝、電器安裝等，也需要在每個指定分包合同中，列出個別的施工規格要求。

Preliminaries

In the common arrangement of the specifications, there is a section on preliminaries preceding the works sections mentioned above. The preliminaries give an introduction to the specifications, explaining the applicability in the contract and the meaning of the terms used. The main part of preliminaries is on the general obligations of the contractor, detailing his or her obligation in (1) taking care of the works and the site; (2) maintaining existing services, features, and trees; (3) maintaining the safety of the workers and the work on site; (4) avoiding any environmental nuisance to the surroundings; and (5) spelling out any special requirements on site personnel, contractor's submissions, materials, and workmanship in general. The preliminaries also specify requirements on any temporary works and services on site, including those of the access roads, site offices, hoardings, and scaffoldings. They also specify the contractor's obligation in providing attendance to all the nominated subcontractors, specialist contractors, government departments and utilities undertakings on site for completion of the work.

初步總覽

在常見的施工規格中，通常會在上述有關建築行業規格的前一章，編寫初步總覽，介紹該施工規格及其適用範圍，以及各名稱的定義。初步總覽的主要部份，是有關承建商一般須遵守的規範，包括：(1) 維護項目工程和工地；(2) 保護現有的公共服務，特別結構和樹木；(3) 保障工人和工程現場的安全；(4) 避免造成任何對周邊環境的滋擾；以及 (5) 指明對工地人員、承建商呈交方案、用料和工藝的任何特殊要求。初步總覽還指定工地上任何臨時工程和裝備的需求，包括那些進出道路、工地辦公室、圍板及棚架。此外，還規定了要協助所有指定分包商、專門承建商、政府部門和公用事業在工地現場上完成合同工程。

Contents of Specifications

In writing the specifications, the following should be specified in each trade: (1) design requirements, (2) materials and submissions, (3) workmanship, and (4) testing and commissioning.

施工規格的內容

編寫施工規格時，各個行業中均需要規定下面各項，包括：(1) 設計要求；(2) 材料及方案呈交；(3) 工藝；(4) 測試和調試。

Design requirements

While the majority of the work is designed by the architect or the engineer, there are some items for which the contractor's design input is required to better suit his or her construction methods or where some specialist supplier's items are being used, such as precast façade, curtain walling, skylight, windows, kitchen cabinet, drywall partition, playground equipment, or shop front. In the specifications, the design parameters and the performance requirements of these items have to be stated. The contractor will be required to submit design drawings and a method statement for the installation for the architect or the engineer's approval.

Some of the design parameters, such as dimensional tolerances in construction, or wind loading requirements, or building services requirements for certain elements, have to be specified in the specifications to govern the quality of the contractor's construction product and the performance requirement during the construction process.

設計要求

雖然工程的主要部份是由建築師或工程師設計的，但亦有少部份項目是需要承建商協助設計，以配合他的施工方法，或是當使用一些專門供應商的產品時，例如：預製外牆、幕牆、天窗、窗戶、廚房櫥櫃、乾牆間隔、遊樂場設備、店面等。這些項目的設計參數和性能要求，都需要在施工規格裏予以說明。承建商亦須就該等項目提交設計圖紙和施工組織設計給建築師或工程師批核。

Table 5-1 Example of writing design requirements for structural concrete work
表5-1 編寫結構性混凝土設計要求的例子

DESIGN REQUIREMENTS
Loadings for design and construction
Design and construct formwork and falsework to withstand the worst combination of the following without causing bulging or deflection:
(i) Total weight of formwork, reinforcement, and wet concrete.
(ii) Construction loads, including dynamic effects of placing, compacting, and construction traffic.
(iii) Wind loads.

Extracted from Architectural Services Department (2012), *General Specification for Building 2012 Edition*. Retrieved 10 June 2017, from https://www.archsd.gov.hk/media/15041/e225.pdf.
摘自建築署(2012)，《2012年版建築物的一般規格》。見https://www.archsd.gov.hk/media/15041/e225.pdf。瀏覽日期：2017年6月10日。

另一方面，一些項目元素的設計參數，如建造時的尺寸寬限，或風力載重，或屋宇裝備的要求，都必須在建築規格中指明，才能有效監管承建商的建築質量和施工表現。

Materials and submissions

In general, only materials for permanent work are specified in the trade sections of the specifications. Materials and tools assistive to the construction or provision of temporary works are deemed to be provided by the contractor and not necessarily specified in the specifications unless in special circumstances. These items shall be included in the preliminaries as mentioned above.

In specifying the building materials, it is necessary to spell out the requirements on: (a) material types, sizes, finishes, performance, and tolerances; (b) material manufacturing process in quality, environmental impact, and safety standards; (c) testing and commissioning of the material to ensure performance quality; (d) packaging and delivery; and (e) warranty of quality. Instead of writing detailed requirements on every material, very often, references are made to some international standards and national standards. The more common standards used in Hong Kong include ISO (International Organization for Standardization), BS (British Standards), ASTM (American Society of Testing and Materials), EN (European Standards), and GB (Guo Biao, Chinese national standards).

Material samples are usually required for the approval by the architect or the engineer before ordering and implementation on site. The material submissions must be accompanied by supporting documents to prove the standard of quality, which include the submission of: (a) shop drawings; (b) material certificates, such as green product certificates; (c) manufacturer's quality compliance certificates, such as ISO 9001, ISO 14001, ISO 50001; and (d) material testing certificates for quality. After the materials have been installed, the architect or the engineer may require the contractor to provide as-built drawings. All these requirements need to be specified in the material section under each trade.

材料及方案呈交

除了在特殊情況下，施工規格一般只會指明會在落成項目中採用的物料，而一般施工時使用的輔助物料和工具，或臨時工程，都是由承建商提供而不用在施工規格中列出。這些項目應列入初步總覽內。

在指定物料時，必須闡明以下要求：(1) 物料類型、飾面、性能、尺寸和寬限；(2) 製造物料過程中的質量、對環境的影響、安全標準；(3) 物料的測試和調試，以確保物料的質量；(4) 包裝和運送；(5) 品質的保證等。很多時候，我們都會引用一些國際標準和國家標準，而不需寫上每一樣材料的具體要求。在香港較常使用的標準包括：ISO（國際標準化組織）、BS（英國標準）、ASTM（美國測試和材料協會）、EN（歐洲標準）和 GB（中國國家標準）。

物料樣品通常需要在物料訂購和現場施工之前，獲得建築師或工程師的批准。提交物料時，必須伴隨著證明文件，以確定其質量標準。當中包括：(1) 製配圖；(2) 材質證書，例如綠色產品證書；(3) 製造商的質量達標證書，如 ISO9001、ISO14001、ISO50001；和 (4) 物料質量的測試證書。物料安裝後，建築師或工程師可能需要承建商提供竣工圖。所有這些要求都必須在各工種的物料部份列明。

Table 5-2 Example of writing specification on materials for timber doors
表5-2 編寫木門物料施工規格的例子

MATERIALS: Timber doors

Doors shall be 45 mm thick hollow or 50 mm solid core flush doors covered with selected hardwood veneer or laminated plastic sheet on both sides and hung to rebated timber frames.

Appropriate number and types of door hinge shall be provided depending on the size and weight of doors used.

When specified, hardwood louvre and frame shall be fitted to the door.

Extracted from Architectural Services Department (2012), *General Specification for Building 2012 Edition*. Retrieved 10 June 2017, from https://www.archsd.gov.hk/media/15041/e225.pdf.
摘自建築署（2012），《2012年版建築物的一般規格》。見https://www.archsd.gov.hk/media/15041/e225.pdf。瀏覽日期：2017年6月10日。

Workmanship

Like building materials, there are many international standards detailing the workmanship requirement for each trade. However, they may not be completely applicable to the local construction industry. Understanding the local trade practice, identifying problems, and finding feasible solutions to achieve the required workmanship demand research and communication with the contractor. This is important to ensure that what is specified can be achievable; otherwise, it is impossible to monitor the quality of workmanship.

In the specifications, requirements on: (1) site preparation and installation work; (2) sample workmanship using mock-ups for approval; (3) installation procedures; (4) quality assurance plan; (5) site safety plan, environmental management plan for the installation work; and sometimes (6) qualification of the site personnel for the proper execution of the work are included in the workmanship section under each trade.

工 藝

每個工種對工藝的要求，如同物料一樣，都有很多國際標準規範。然而，它們未必完全適用於本地的建築行業。要達到所需的工藝要求，我們必須透過研究和與承建商溝通，並明白本地行業的慣常操作，了解問題所在並找出可行的解決方案。重要的是規格裏的要求，必須是可以操作的，否則，便無從監督工藝的質量。

在施工規格上，要列明對各工種工藝的要求，包括：(1) 現場準備和安裝工作；(2) 利用施工樣板作為工藝樣品的審批；(3) 安裝程序；(4) 質量保證方案；(5) 施工時的工地安全計劃、環境管理計劃；及有些時候，(6) 適合執行合同工程人員的資歷。

Table 5-3 Example of writing specification on workmanship for door frames
表5-3：編寫木門框工藝施工規格的例子

WORKMANSHIP: Door Frames
Timber door frames shall be securely fixed to the partition framework with diagonal braces on each side to stabilize the mid-height point of the frame.
Door frames at corners and tee junctions shall be positioned to allow sufficient clearance between the back of the door and the intersecting partition to allow the mounting of a surface-mounted door closer and enable the door to open a full 90 degrees without the closer body striking the intersecting partition face.

Extracted from Architectural Services Department (2012), *General Specification for Building 2012 Edition*. Retrieved 10 June 2017, from https://www.archsd.gov.hk/media/15041/e225.pdf.
摘自建築署（2012），《2012年版建築物的一般規格》。見https://www.archsd.gov.hk/media/15041/e225.pdf。瀏覽日期：2017年6月10日。

Table 5-4 Example of writing specification on testing requirement of tile grouting
表5-4：編寫測試瓷磚填縫的例子

TESTING: Tile grouting			
The quality tests for tile grouting shall be as follows:			
Test items	Test method	Acceptance standards	Remarks
1. Linear shrinkage	ANSI A–108/A1 18/ A136.1-2011	1 day shrinkage < 0.1% 7 days shrinkage < 0.2%	Cast and store grout specimens at 21°–25°C, 45–55% R.H.
2. Water absorption	ANSI A108/A13 601-2011	From 50% R.H. to immersion < 5% From immersion to dry < 7%	Determine water absorption from 50% R.H. to immersion and from immersion to dry

Extracted from Architectural Services Department (2012), *General Specification for Building 2012 Edition*. Retrieved 10 June 2017, from https://www.archsd.gov.hk/media/15041/e225.pdf.
摘自建築署（2012），《2012年版建築物的一般規格》。見https://www.archsd.gov.hk/media/15041/e225.pdf。瀏覽日期：2017年6月10日。

Testing and commissioning

Testing and commissioning are important for quality assurance of the work. All quality testing and commissioning testing requirements, including the number of sampling tests, the type of testing required, detailed procedures of each test, requirements on third-party testing institutes, must be specified under the testing and commissioning section of each trade.

In addition, consideration should be given to specifying the criteria for any necessary retesting and remedial work required when the tests fail. The number of samples for retesting, procedures, and penalty should also be specified.

測試及調試

測試及調試是工程中重要的質量保證。所有的質量測試和調試要求，包括取樣測試量、所需的測試類型、每個測試的詳細過程、對第三方測試機構的要求等，都需要在各個工種的測試與調試中規定。

除了列明測試與調試的要求外，也應考慮加入當測試失敗後所需要的重新測試和善後工作；還應該指明重新測試樣本的數量、程序和處罰條款。

Writing Specifications

Since specifications are part of the contract document, the language used must be concise and precise. It should be clearly understandable, without ambiguity, and well-coordinated with the drawings, BQs, and other contract documents. All terms used and references made should be consistent throughout the specifications and contract documents. As the purpose of the specifications is to inform the contractor about the end product which the architect or engineer designs, it is addressed to the contractor only. Other items, or issues, although within the project but outside the scope of the work, such as the work of the nominated subcontractors, should not be specified in the contract, apart from drawing the contractor's attention to their existence. In specifying materials, research should be carried out to ensure the material is available in the market with enough suppliers to avoid monopoly. In specifying workmanship, it is advisable to have a dialogue with the local industry to work out an implementable way of installation before putting it in the specifications. If we specify material and workmanship making reference to international standards, it is necessary to check that acceptance testing equipment and laboratories are easily available in the industry. Any requirement on materials and workmanship that cannot be verified for their acceptance will not be enforceable under the contract and therefore should not be included in the specifications. There shall also be reasonable penalty clauses for any substandard material and workmanship, to ensure quality of work from the contractor. However, and of equal importance, there shall be bonus clauses to encourage the contractor to make innovative contributions to the contract. After all, a happy contractor will deliver a good project.

編寫施工規格

由於施工規格是合同文件的一部份，所以使用的語言必須簡潔和準確，易於明白，毫不含糊，並與建築圖紙、工程量清單以及其他合同文件一致。所有條款、參考資料，應該在整個規格和合同文件內統一。由於施工規格的目的，是建築師或工程師用來告知承建商其對建築設計的要求，故此，只應針對承建商而寫。其他項目或議題，儘管可能是項目以內，但是在合同工作範圍以外的，如指定分包商的工作，除了讓承建商知道他們是與合同工程有關外，其餘的都不應該在施工規格中出現。在指定物料時，應進行研究，確保該物料可在市場上有足夠的供應，以避免壟斷。在指定工藝時，最好先與本地業界溝通，商討切實可行的方案，才把它制定在規格裏。如果我們是參照國際標準來指定物料和工藝，務必確定驗收測試的設備和實驗室能在行業內容易找到。任何物料和工藝的要求，如不能核實他們的質量，將會無法按合同執行監管，而這些不能監管的要求，都不應納入施工規格內。對任何不合標準的物料和工藝，應適量加入處罰條款，以確保承建商的工程質量。但同樣重要的是，施工規格中應該設有獎勵條款，鼓勵承建商對工程合同作出創新貢獻。畢竟，一個愜意的承建商更能完成一個優質的工程項目。

About the Authors

WONG Wah Sang (黃華生) is an experienced architect and has taught building technology in the Department of Architecture, the University of Hong Kong, for 25 years. He is an expert in related law and local practices and is in charge of the professional assessment for architects in the subject of building technology and materials. He serves on the editorial board of several international journals and has lectured and written on various areas of architecture for international conferences and local architectural education institutions. He has also served in different positions in the Hong Kong Institute of Architects and Architects Registration Board as well as some government committees on building matters.

CHAN Wing Yan, Alice (陳詠欣) is a registered architect in Hong Kong. She established her own private practice, Architectural Project Unit Limited, in 2012. Her specialty in architecture includes renovation of shopping malls, building façade design and repairs, and villa design. She is also a part-time lecturer in the Department of Architecture, the University of Hong Kong.

WAI Chui Chi, Rosman (衞翠芷) is a registered architect who specializes in specification writing and has much experience in Hong Kong public housing. In addition to her professional degree in architecture, she has a master's degree in architectural conservation (with distinction) and a PhD from the University of Hong Kong. She has presented at universities and at local and international conferences on various subjects in architecture. She serves on the Architects Registration Board (2014–2018) and was the vice president of the Hong Kong Institute of Architects (2015–2016).

KEE Yee Chun, Tris (祁宜臻) is a registered architect in Hong Kong. She is an authorized person and an associate professor in the Faculty of Design and Environment at the Technology and Higher Education Institute of Hong Kong. Tris has taught professional practice, contract, design studio, and building technology. She has lectured and written about architecture in Hong Kong and overseas. Devoting her time to serve the architecture community, Tris is also the editor-in-chief of the *Hong Kong Institute of Architects Journal*, and an executive committee member in the Hong Kong Interior Design Association and Hong Kong Architecture Centre.

Copy-editor

Vivian AI 艾丹丹 (MCIM, MIPRA) is a Chinese Australian, who is professionally trained as an architect and possesses extensive experience in integrated marketing, branding, and digital communication gained through her management involvement in Big Four, listed financial groups, and international design firms. As the founder and editor-in-chief of *City Outlook*, a charity initiative monthly bilingual design publication, Vivian voluntarily devotes much of her time and energy contributing to the wider Hong Kong community using her knowledge and network across multiple sectors.